Interactive and Animated Cartography

**Prentice Hall Series
in Geographic
Information Science**

KEITH C. CLARKE,
Series Editor

Interactive and Animated Cartography

Michael P. Peterson
University of Nebraska at Omaha

Prentice Hall, Englewood Cliffs, New Jersey 07632

Library of Congress Cataloging-in-Publication Data

Peterson, Michael P.
   Interactive and animated cartography / Michael P. Peterson.
      p.   cm.
   Includes bibliographical references and index.
   ISBN: 0-13-079104-0
   1. Cartography--Data processing. 2. Interactive computer systems.
   I. Title.
   GA 102.4.E4P48 1995
   526'.0285--dc20   94-20878
                                                                         CIP

In Memory of Barbara B. Petchenik

Acquisitions editor*: Ray Henderson*
Assistant editor: *Wendy Rivers*
Project manager*: Robert C. Walters, PMI*
Cover designer: *Pat Woczyk*
Production coordinator*: Trudy Pisciotti*

 © 1995 by Prentice-Hall, Inc.
A Simon & Schuster Company
Englewood Cliffs, New Jersey 07632

Printed in the United States of America

10   9   8   7   6   5   4   3   2   1

ISBN: 0-13-079104-0

PRENTICE-HALL INTERNATIONAL (UK) LIMITED, LONDON
PRENTICE-HALL OF AUSTRALIA PTY. LIMITED, SYDNEY
PRENTICE-HALL CANADA INC. TORONTO
PRENTICE-HALL HISPANOAMERICANA, S.A., MEXICO
PRENTICE-HALL OF INDIA PRIVATE LIMITED, NEW DELHI
PRENTICE-HALL OF JAPAN, INC., TOKYO
SIMON & SCHUSTER ASIA PTE. LTD., SINGAPORE
EDITORA PRENTICE-HALL DO BRASIL, LTDA., RIO DE JANEIRO

Contents

Foreword

In methodical detail *Interactive and Animated Cartography* explores an innovative technology that can free mapping from the expediency of static images and afford thorough, fully engaging cartographic treatments of geography, history, and earth science. Michael Peterson paints a vivid picture of a New Cartography in which maps narrate, cartographic symbols move, and displays cough up relevant information when the viewer pokes them in the right places. In rescuing both makers and users of maps from the straitjacket of ossified single-map solutions, interactive mapping promises a cartographic revolution as sweeping in its effects as the replacement of copyists by printers in the late fifteenth and early sixteenth centuries.

Two emerging consequences of cartography's electronic revolution are substantial alterations in the nature of maps and the nature of map making. No longer merely an object or artifact, the visual map is swiftly becoming an interface between cartographic data and the map user. This shift, in turn, has blurred traditional distinctions between map maker and map user. After all, the user who can probe the data and create customized displays is both a user and a maker of maps whom I suppose we ought to call, quite simply, a mapper. Moreover, secondary viewers who can modify or look beyond this first mapper's authored representations clearly are more empowered than the traditional map user who can like it or lump it but can't change it. They too are mappers.

Interactive cartography also promises a renewed appreciation of geography, which means, quite literally, "writing about the Earth." Geographic writing was never easy because words arranged in one-dimensional sentences and paragraphs are seldom adequate for describing and explaining phenomena that occur in a three-dimensional space and evolved over time. A linear narrative adapted to one person's interest and knowledge will usually bore some readers and confuse others. Cost, aesthetics, and page format limit an author's ability to integrate words, maps, photographs, and other images essential to spatial understanding. But as Peterson illustrates, geographic representations can escape the tyranny of the printed page. When interactive cartography and multimedia fully mature, readers can proceed at their own pace, call for graphic narratives customized to their unique interests and experience, and navigate freely through richly rewarding electronic atlases.

While it would be folly to forecast the demise of the paper map — as the pages of this book demonstrate, paper can be enormously user-friendly — animated and interactive maps will have a pronounced impact on their hard-copy counterparts. One effect will be increased recognition among map viewers that the static map is but one of many cartographic views of the same phenomenon or data set. Another consequence will be paper maps designed to summarize an animated presentation or interactive work session. Ironically, instead of displacing paper maps, computerized cartography seems likely to make them more common. The mass-market printed map might be at risk, but the one-off cartographic summary custom tailored by the mapper to his or her unique interests is on the rise.

Cartography is ripe for a book like this. In the decade since the first Macintosh, computers have become smaller, faster, and cheaper. Equally significant, the electronic storage needed for massive cartographic data bases has become more spacious, more quickly accessible, and less expensive. Building on this improved hardware, government agencies and commercial mapping firms have compiled a rich array of electronic cartographic data, and developers of mapping software have moved beyond static graphics to animation and interaction. Keeping pace, cartographic researchers have developed theory and techniques that adapt contemporary technology and anticipate further progress. As Peterson explains, although guidelines for designing static maps are often relevant to dynamic maps, human interaction and visual variables based on motion and time require radically new conceptual frameworks and practical guidelines.

An accessible introduction to timely new ideas, *Interactive and Animated Cartography* also offers experience-tested insights about why maps work and how to design them. Concise accounts of recent advances in map-making technology provide terminology and concepts to help the reader understand and start using the New Cartography.

Mark Monmonier
Syracuse, New York

Preface

In the Fall of 1985, I completed a mapping program for the Apple Macintosh computer, a computer first released only 18 months earlier. The program produced choropleth maps, a type of map that uses shadings to depict data values for areas such as states or counties. It placed the map in the memory of the computer as individual polygon "objects" based on a file of x, y coordinates. Because the map resided in memory, it could be quickly updated with new shadings to depict another variable or another method of data classification.

While the Macintosh computer is fairly easy to use, programming the computer is another matter. I often worked past midnight to incorporate the various user interface elements. Early one morning I successfully implemented a menu with a variety of classification options and different numbers of classes from 2 to 16. Each menu selection created a new map on the screen in less than three seconds. Just ten years before, in an undergraduate course, I spent over twenty hours constructing just one such map. A map that I created in 1977 with an IBM 370/158 mainframe computer took more than an hour to complete. It required the use of computer cards and a tape-drive to run a pen plotter that drew the map with crossed-line shadings. Using a DEC VAX minicomputer in 1982, five minutes were required to create a crude-looking map on the screen of a low-resolution video terminal. Now I had placed the map on a high-resolution screen of a microcomputer in under three seconds. I spent the remainder of those early morning hours testing each of the classification options and observing with some astonishment how quickly the maps were produced.

As I tested the program, it became clear to me what the computer really meant to cartography. It was not merely a tool to speed the creation of maps on paper. The computer represented a different way of viewing and interacting with maps. It was apparent to me that the human relationship with maps was entering a new era.

Although I didn't realize it at the time, this book had its beginnings on that early morning. I continued programming for some time, eventually achieving a map animation display rate of sixty maps a second. Convinced of the importance of both interaction and animation in the display of maps, I began to examine the importance of the medium of paper in cartography. It seemed that the medium had defined cartography by influencing the way maps look and the way they are used. The static nature of the paper map was the limiting factor in improving the human interface with maps. It handicapped our minds and limited our understanding of the physical and human environment. Worst of all, the computer continued to be used to create maps on paper — maps that duplicated in exacting detail the limitations of the hand-drawn printed map. The computer's potential for interaction and animation was essentially ignored.

This book is about adding interaction and animation to the display of maps. Chapter 1 reviews the related developments in cartography. Chapter 2 presents a theoretical background for how maps communicate. Chapter 3 introduces the potential of interaction and animation with maps. Chapters 4 through 9 examine the tools that are available for interactive and animated cartography. Chapters 10 and 11 contain more specific examples of applications of this software. Concluding comments are made in Chapter 12. Appendix A introduces programming in Fortran and C. Commercially available software are listed in Appendix B. Appendix C describes Internet and the resources that are available through this service. Appendix D lists other sources of software and Appendix E provides sources of maps and imagery.

This book is intended for the two major types of microcomputers: the Apple Macintosh and Intel-based PC computers with the Microsoft Windows operating system. The point-and-click operating systems of the Macintosh and Windows leads to a greater degree of user interaction with programs. An emphasis is placed on the Apple Macintosh because it is widely used in cartography. However, many of the Macintosh programs that are mentioned in this book are now also available for Microsoft Windows. Windows has many similarities to the Macintosh user interface, including menus, windows and dialogs and support for a mouse pointing device. With minimal computer background and a small amount of training in the Macintosh or the Windows operating system, you should be able to complete the exercises at the end of each chapter.

A number of people have helped me with this book and I am indebted to them all. Keith Clarke, David DiBiase, Richard Crooker, Mark Monmonier, and Phillip Muehrcke provided valuable comments on earlier drafts of the manuscript. All of the students in my 1993 seminar on Computer Mapping and Data Analysis offered insightful suggestions, including Bernadette Doerr, Roy Hopkins, Ed Kelley, Robert Tusa, Dwight Warak, LuAnn Wentworth, Jay Parsely, and Charles St. Lucas. This work was begun under a Fulbright Fellowship to Germany. I would like to thank the Fulbright Scholar Program, the German Fulbright Commission, and Prof. Dr. Ulrich Freitag for extending an invitation to the Free University of Berlin. Doris Dransch, a doctoral candidate at the

Free University, offered valuable comments on the initial versions and expressed an interest in the progress of the book throughout. My immediate colleagues — Marvin Barton, Charles Gildersleeve, Roger Hubbard, Jeffrey Peake, and Robert Shuster — provided a combination of information, encouragement and support throughout this project. Ray Henderson and Wendy Rivers of Prentice Hall, and Robert Walters of Prepress Management need to be thanked for getting this book to press. Finally, I owe much to my family. To my two daughters, Sarah and Amelia, now 12 and 7, I wish to say thank you for understanding how much time goes into writing a book. A special thank you goes to my wife, Kathy, who was not only extremely supportive throughout this project but also provided many valuable suggestions on the book, its organization, wording, and style of presentation.

Michael P. Peterson
Omaha, Nebraska

Part I

MAPS, MEDIUM AND THE MIND

A medium is the carrier of information. It is used to transmit knowledge and ideas between people. Each medium has a certain potential for communication.

Maps, as depictions of the world, represent the way we communicate about the world around us. For centuries maps have been constructed on the medium of paper. The computer has been used for many years to assist the cartographer in making maps on paper. Now the computer is being used directly for the display of maps. For cartography the computer medium presents the potential of interaction and animation.

Part I of this book creates a framework for a more interactive and animated cartography. The first chapter presents an overview of cartography and examines the influence of the paper medium in cartography. Chapter 2 looks at how maps communicate information. Chapter 3 examines some of the possibilities of interaction and animation for the display of maps.

1

Maps and the Changing Medium

1.1 INTRODUCTION

Maps are abstractions of the world that help us to understand our environment. Their purpose is to communicate information. For many people, maps do not appear to achieve this objective. Most people have poorly developed mental representations both of the location of places and the distribution of phenomena. While cartographers attempt to create informative and visually pleasing graphic displays, it appears that the maps they produce often fail to communicate a meaningful and lasting impression.

Maps are an indispensable form of information display. They represent a vital link between people and the world around them. However, it is estimated that over half of the population avoid using maps entirely. The results of tests given to freshman university students indicate an inadequate knowledge of the location of even the most basic places. Indeed, many could not find the United States on a world map. While the educational system may be blamed for not specifically teaching about maps or their use, it may be maps themselves that are largely to blame.

Until recently, maps were exclusively designed and constructed to be printed on paper. The printing of maps on paper was economical, and this facilitated their distribution. But paper maps may not represent the best way to convey information. Not only is a dynamic and complex world depicted with static representations, but the need to create individual maps requires that critical choices be made concerning the method of representation, including such factors as the amount of generalization, the method of data classification, and symbolization. The map user has no control over the information that is presented on a map or how it is displayed.

Technology is making it possible to "re-think" how maps are presented. In particular, computers are making more interactive and dynamic methods of map display possible. Microcomputers are making this technology more generally available. It is now possible to restructure the human relationship with maps — to change how maps are presented and how they are used.

This book is about the human interface with maps and how the computer can be used to create a more dynamic form of map use. It is about how we can change the human interaction with maps and how we can ultimately improve people's conception of the world in which they live.

1.2 COMMUNICATION IN CARTOGRAPHY

A concern with improving the map as a form of communication is not new. The goal of the cartographer was always the creation of a usable product. Since the 1950s, however, the process of map communication has been examined more explicitly. The initial aim of this emphasis in *cartographic communication*, as defined by Robinson (1952), was to determine the user response to individual map design elements with the goal of modifying the map design process. The concept of the map as communication device has since transformed cartography. Instead of being only concerned with the techniques of map construction and an endless adaptation to technological innovations, cartographers could legitimately examine the function and purpose of maps. Maps themselves became the object of philosophical and scientific inquiry. As a result, cartography was transformed into "the science of communicating information between individuals by the use of a map" (Morrison 1978, 97).

The specific research in cartographic communication had two major phases: (1) Initially concerned with the stimulus-response relationship of individual symbols, cartography incorporated the research methods of psychophysics and attempted to determine how to perceptually adjust the scaling of symbols, and (2) Beginning in the late 1970s, cartographic research developed in the direction of cognitive psychology and became concerned with how maps were processed and remembered. The goal, however, remained to find a way to improve maps so that they could communicate more, or more accurate, information to the map user.

Research in cartographic communication has been guided by general models of communication. In this view cartographic communication is defined as the process by which information is selected, symbolized on a map, and subsequently perceived, recognized, and interpreted by the map user (Wood 1972, 123). The role of the cartographer is to create a depiction of the world that coincides with some part of the map user's reality. Models of cartographic communication divide the process into separate stages (Figure 1.1). The stages can be placed into the cartographer's domain or the map user's domain; the two domains coming together with the map. In this conception, the cartographer controls the communication process.

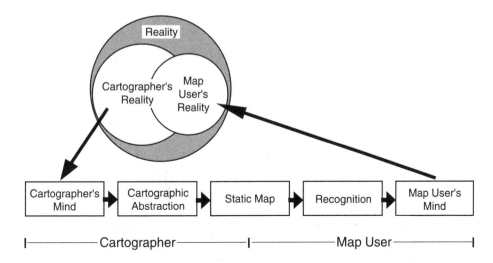

Figure 1.1   A model of cartographic communication.   In this model the map user must adjust to the depiction of the world that is created by the cartographer; the cartographer controls the communication process.

As modeled, the cartographer transforms his reality — the message — onto paper — the medium. If possible, the map reader extracts the message from the medium, and communication occurs if there is an overlap between the two separate "realities." Communication models also incorporated the concept of "noise," analogous to electronic interference, on each of the links. The noise interferes with the message and has a cumulative, negative effect across the different links.

This general conception of the cartographic communication process is changing. The map is no longer the static element in the communication chain. The map is being transformed into a more interactive, user-controlled type of display. As the cartographer is no longer constrained by the limitations of creating a single map, so the map user is given more possibilities in interacting with the map display, going beyond the limitations of the static map. The model presented in Figure 1.2 incorporates interaction in the use of maps.   The interactive map is a product of a feedback loop, a response mechanism implemented within the user interface that makes interaction with the map possible. The interactive map provides the map user more control over the communication process.

The overall conception of cartographic communication has also been based on the concept of the *average map user*. It has long been recognized that this concept is a fallacy and yet maps are still designed to appeal to the cartographer's conception of such a map user. A more interactive mapping environment makes it possible for users to adjust the map display according to their needs and abilities.

The animation of maps will add yet another element to our conception of cartographic communication. There is evidence to indicate that the human mind is

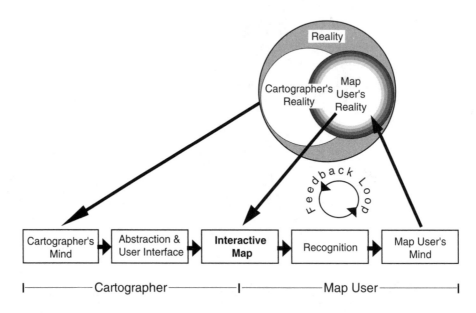

Figure 1.2   Model of cartographic communication incorporating interaction. In this model the cartographer provides a general map use environment but the map user controls what is depicted and how.

extremely adept at gathering information from moving displays. Indeed, our ability to detect change in a visual display may be more developed than the recognition of static images.

1.3 CARTOGRAPHIC COMPUTERIZATION

Beginning in the early 1960s and concurrent with research in cartographic communication, cartography was greatly influenced by computer technology. The adaptation to this new tool occurred very quickly, even though early methods of graphic output produced crude depictions. Programs such as SYMAP created maps using the typewriter-like characters of the computer line-printer. By the early 1970s coordinate plotters were more common and made the emulation of hand drafting possible. More sophisticated output devices, some incorporating color, became available during the 1980s, and programs were written to take advantage of the capabilities of these graphics peripherals.

An interest in more dynamic methods of mapping is a result of several trends in information management that have had a strong influence on cartography, including GIS, scientific visualization, and virtual reality. Many of these developments are in turn related to the miniaturization of computers, first with minicomputers in the late 1960s and then with microcomputers a decade later. The result was the development of less expensive and more interactive computers.

1.3.1 Geographic Information Systems

Geographic Information Systems (GIS) have had a major influence on cartography since the late 1970s. Such systems are used for the management and analysis of various types of geographic information and incorporate the functions of data capture, storage, retrieval, analysis and display. GISs are used for a wide variety of applications and many include sophisticated query and output capabilities. GISs have prompted a major demand for computer-assisted methods of mapping and are beginning to stimulate new methods of map display.

The major emphasis of a GIS is data manipulation. The cost of GIS software and hardware is generally fairly high. In addition, the systems are complicated. The time it takes to learn a single software package can be greater than two years of daily use Because such systems are generally expensive, difficult to use and designed for the creation of maps for specific resource or demographic analysis purposes, they can be viewed as a type of "cartography for the few."

GIS programs for microcomputers are becoming less expensive and easier to use. A new development related to GIS is the availability of "viewers," specialized programs for microcomputers that allow the user to interactively query an existing GIS data base through a point-and-click user interface (e.g., ESRI's ArcView). The programs require a minimal amount of training, thereby bringing the technology to a wider audience. Viewers are "empowering" the GIS user by providing more intuitive tools to investigate data and create their own map displays.

1.3.2 Visualization

Visualization is the creation of computer graphics images that display data for human interpretation, particularly of multidimensional scientific data. It has been interpreted broadly as a method of computing that incorporates data collection, organization, modeling, and representation. Visualization is based on the human ability to impose order and identify patterns.

An outgrowth of statistical analysis, visualization is now used in a variety of disciplines. It has strongly influenced all forms of data analysis, and the techniques are beginning to be incorporated in cartography. Important elements of the visualization interface are interactivity and animation. For example, programs are available to graph x, y, and z variables (e.g., Abacus MacSpin). The programs use animation to "spin" a three-dimensional graph depicting the data. The animation makes a cloud of points visible in two-dimensions that exist in three-dimensions.

Cartographic visualization, sometimes called geographic visualization, is interpreted as the use of similar techniques for the display of maps. In discussing the possibilities of computer technology, MacEachren and Monmonier write:

> The computer facilitates direct depiction of movement and change, multiple views of the same data, user interaction with maps, realism (through three-dimensional stereo views and other techniques), false realism (through fractal generation of landscapes), and the mixing of maps with other graphics, text, and sound. Geographic visualization using our growing array of computer technology allows visual thinking/map interaction to proceed in real time with cartographic displays presented as quickly as an analyst can think of the need for them (1992, 197).

The strong interest in visualization within cartography is partly a response to a data-base view of maps that emerged along with the growth of geographic information systems. Implicit within the data-base view is the notion that all elements of a map can be decomposed and represented within a file on the computer. Further, once so encoded, all analysis can proceed within the data base without any need for graphical representation or human involvement. This view of maps can be termed "nongraphical cartography."

Visualization reaffirms the importance of the graphical illustration in all aspects of analysis and interpretation. It recognizes that the human has special abilities to interpret graphical displays and that these abilities must be utilized.

1.3.3 Virtual Reality

Virtual reality is a step beyond visualization. It is a simulation that includes stereo vision, stereo sound and the ability to control the display, often through the use of a special "data glove." The observed scene may also be coordinated with head movements. Virtual reality games have become a major entertainment attraction. An important aspect of virtual reality is that the visual and auditory senses are flooded to such an extent that the individual "loses touch" with the current reality and has the feeling of entering another place.

Virtual reality is moving from entertainment into the mainstream of information technology. It is being used for engineering, aerospace, and industrial applications. The technology would seem to have a number of uses in cartography. It could be used to explore the area represented by a map. People could be made to feel as if they are walking across a map and then "fall into" the reality it represents. Then users could climb back onto the map and "walk" to another place, even one thousands of miles away.

The potential of virtual reality in cartography is to bridge the gap between map and reality. It can define the relationship between the abstract symbols on a map and what they represent. Virtual reality holds the promise of taking a person into the world that is depicted on a map.

1.3.4 Technology and Communication

Until recently the computer has been used primarily to automate the production of maps on paper. The computer was viewed as a tool to make the creation of maps easier for the cartographer and possible for the noncartographer. This view of the computer is changing. Computer mapping software is beginning to incorporate an interactive, in some cases animated, form of cartography — a type of map use that is not possible with the printed map. The computer is being used not only as a tool to help make maps on paper or search a data-base but as a medium of communication.

Since the 1950s, cartography has proceeded in two essentially distinct directions — communication and computerization. With the miniaturization of computers, these two directions are merging. The computer is beginning to be used to enhance the effectiveness of maps. The goal is to make maps more useful and accessible than they have previously been with paper. With interactive cartography the map user stands in the middle of a map creation and map-use process and controls what is depicted and how. The animation of maps makes the depiction of time and other variables possible. The goal of both interactive and animated cartography, like that of cartographic communication, is to improve the effectiveness of maps as a form of communication.

1.4 A PERSPECTIVE ON MAPS

In making the transition to this new era of map use we need to place maps and the medium of cartography in a historical perspective. How did cartography begin? How has cartography been influenced by the paper medium?

1.4.1 The First Map

It has been a long time since the creation of the first map — so long that we can only guess when and how it might have occurred. Maps may have been used initially for way-finding or to delineate territory. A possible "first map" scenario might have two early humans meeting on a beach; one is familiar with the area and the other is a newcomer. In the process of trying to explain where food and water could be found or where the dangerous areas are, the first human, in a moment of frustration in not being able to describe the location of these places through verbal communication or with gestures, reaches for a stick and draws some lines and shapes in the sand. While many would argue that this action represented an intellectual leap for mankind, of greater importance would have been that moment when the second individual, after examining the lines and shapes in the sand, nodded his head in understanding. For it was at this point that the depiction took on meaning — indeed, that it became a map.

This map in the sand was not the kind of static map that we find today on paper. The meaning of the symbols would have been explained as the map was created. The second person would have asked questions that influenced how the map was drawn. As new features were added, they may have obscured part of the existing map. Indeed this first map would have been very much like the kind of interactive and dynamic map that we are attempting to create today with the computer.

In the intervening years the map has been transformed into the static object it is today. Early maps were carved in stone. Later, paper became the predominant medium for the distribution of maps.

1.4.2 The Mental Map

Before any map was drawn in the sand, it existed first in the mind of an individual. The term "mental map" is used to describe an internal representation that is similar to a map but arises from memory. It is a useful concept for understanding what was communicated with the drawing in the sand. In the example above, the map in the sand can be viewed as an outward expression of a mental map that was formed in the mind of the first person as a result of personal experience in the area. After examining the map in the sand, the second person also had a mental map of the area based on the map in the sand.

There is an obvious qualitative difference between the two mental maps based on how they were created. Although we all still have mental maps based on personal experience that help guide us through our daily environment, many if not most of our mental maps are derived from external maps. A conception of the outline of a state would be an example of a mental map derived from a printed map. Even if we were able to fly over the country at a sufficient altitude to view an entire state, it would not be outlined as it is on a map. Another example of a map-based mental map would be a conception of the distribution of population concentrations or intangibles such as income or cancer mortality. These distributions can only take form on a map.

Mental maps exhibit distortions and biases. Distances are often incorrect. The extent of these distortions is difficult to measure because we are not able to adequately externalize these maps. We know this because people are rarely satisfied with the mental maps they have drawn. The ability to see our own errors in the maps we have drawn indicates that such externalized maps do not adequately reflect the "map in the head." The exact form and level of detail of mental maps is difficult to determine.

Mental maps can also embody emotion. Downs and Stea (1973) show how people in an urban area have a mental map that reflects perceived areas of crime. Fear is just one of many emotions that can be encoded on mental maps.

The mental conceptions that we have of the world around us play a large role in our relationship with the environment, our general actions, and our behavior. These mental maps are our connections to the "real" world. People with a poorly developed mental map have a very incomplete picture of their surroundings and may feel isolated and fearful of the "outside" world. Mental maps not only provide a better visual conception of our surroundings but also shape our attitudes about people and places (see Tuan 1974 on the perception of place).

1.4.3 Origins of Cartography

The map in the sand would have represented an initial formalization of a mental map. The process of externalizing such representations in the form of maps is the basis of cartography. Surveying techniques, ultimately aerial photography and satellite imagery, were incorporated to create a view of the earth beyond the perception of an individual. Today maps have become such a critically important aspect of our society that we can't imagine a world without them. Every person is in some way affected by maps whether they use them directly or not, from cadastral maps that delimit property boundaries to the maps used by the military to assure our security.

Cartographers have been relegated to the task of constructing maps, a complicated process of abstracting the "real world" to create a generalized yet useful, visual depiction of our environment. The efforts of many cartographers in the past 300 years have concentrated on ways of accurately transforming the round earth to a flat medium. A variety of map projection methods were devised during this period. None perfectly depicted all spherical properties, although many were acceptable for different purposes. The efforts of these early cartographers made it possible to use the flat medium of paper to depict large parts of the earth's spherical surface.

Printing solved another obstacle in cartography, that of map distribution. Before printing was invented, each map had to be faithfully duplicated by hand, as was often done by monks in the Middle Ages. Very few individuals at this time had the opportunity to view maps, and the ownership of a map was economically limited. The printing of maps was first accomplished in 1472 in Augsburg (Germany) with a woodcut (Bagrow and Skelton 1964). Equally important to cartography was the development of other printing technologies, including the copper plate that dominated printing from about 1650 to 1850 (Robinson 1982, 22). Rapid changes in printing took place in the second half of the nineteenth century with the introduction of lithography. Cartography has steadily evolved with changes in printing technology.

1.4.4 Maps and Society

While the printing of maps increased the potential for map use, the information that they contained was often considered to be secret for military or economic reasons. Maps were used initially by people to gain advantage over others, particularly through military means. The history of cartography shows great technical advances during the time of war. Obviously, many people in power recognized the utility of maps, and it was this utility for specific purposes that limited their general use.

Cartographers have examined the close relationship between the development of maps and the rise of nation-states. In *The Power of Maps*, Wood (1992, 43) argues that the map can be viewed as an instrument of the nation-state to wage war, to assess taxes, and to exploit strategic resources. According to Harley (1988), the nation-state is mostly interested in stability and longevity. To this end cartography is "primarily a form of political discourse concerned with the acquisition and maintenance of power" (Harley, quoted in Wood 1992, 43).

The role of maps in maintaining power is particularly evident in totalitarian governments. The former German Democratic Republic, for example, undertook the systematic distortion of large-scale maps so that they could not be used against the country in time of war. This type of intentional distortion of maps was common in the Soviet Union and many eastern European countries. With the opening of these countries, stories are now being told of the great difficulties cartographers had in creating false maps, particularly the updating of these maps with new features. Complicated procedures had to be devised to place new items relative to the false position of existing map elements.

Even democratic governments like the United States classify as top secret a great many maps. Much of the information collected and mapped by various government agencies in the United States, especially the military, is never made public. Companies also collect demographic data that is mapped but kept secret for reasons of competition.

Governments and mapping companies publish a great many maps. This general availability of maps has only been a phenomenon of the last 100 to 200 years, predominantly in the Western world. The increased access to maps coincided with the creation of more democratic societies. This relationship is not coincidental. Many would argue that the formation and maintenance of democratic societies is dependent upon access to information. Without effective means of distributing information, democratic societies fall because people cannot make informed decisions. As maps have been used by totalitarian governments to maintain their power, maps can serve an important role in providing information to people in order to preserve democracies. The creation of more effective methods of map presentation and map use helps to communicate information to maintain an informed society.

1.4.5 The Two Types of Maps

Most people think of maps as simply depicting the location of places. Important as they are in creating a conception of where things are in relation to each other, *general reference maps* give an incomplete depiction of the earth. The type of map that has more influence on our understanding of world distributions are maps that emphasize a particular phenomenon. The *thematic map* is used to depict such distributions as population, agricultural practices, or landscape types. The purpose of these maps, whose origins go back to the latter half of the seventeenth century (Robinson 1982, 17), is to show "distribution-as-thing." In contrast to the general reference maps that emphasize location and are used for navigation, thematic maps concentrate on the depiction of single or multiple related phenomena. Petchenik (1978) describes the distinction as "in place, about space." General reference maps put objects in place, whereas thematic maps inform about distributions in space. The growth of thematic cartography during this century has been particularly dramatic.

1.4.6 Map Symbols

Three basic types of symbols are used on both general reference and thematic maps. These symbol types are *point, line,* and *area.* Point symbols depict information at points (Figure 1.3). Two common examples of point symbol maps are the dot map and the graduated circle map. On dot maps the number of dots is increased to show increasing quantity. With graduated symbol maps the symbol size is increased to show greater quantity. There are two types of point symbols: (1) geometric symbols, such as the dot, circle, or square, which bear no particular resemblance to what is being mapped, and (2) pictographic point symbols, which attempt to create a graphic association between the symbol and the mapped variable.

A line symbol map represents quantity along a line by assigning different values to lines or by varying the width of the line. On an isarithmic map different values are assigned to the lines (Figure 1.4). The isarithmic map comes in many forms. Contour maps that depict elevations and isotherm maps that show temperature are both examples of this mapping method. The width of the line is varied on a graduated line map to represent value, such as the flow of a river or traffic on a highway.

Area symbols depict data for areas such as polygons. Examples include the choropleth and cartogram maps (Figure 1.5). The choropleth mapping method is commonly used for depicting socioeconomic data, partly because this type of map can be

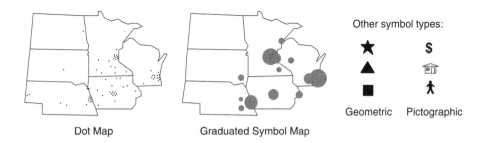

Figure 1.3 Examples of point symbol maps. The dot map depicts quantity by increasing the number of dots. The graduated symbol map varies the size of the symbol. Point symbols can be either geometric in shape (circle, square) or pictographic.

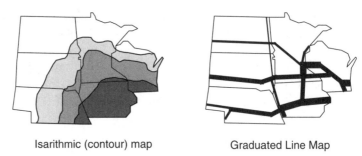

Isarithmic (contour) map   Graduated Line Map

Figure 1.4   Examples of line symbol maps. The isarithmic map has lines drawn through points of equal value. The shaded form of the isarithmic map is shown here. The graduated line map increases in width to show greater quantity, usually some type of flow.

easily created by computer. On the cartogram the size of the enumeration unit is changed to reflect its value. Cartograms come in two variations. Figure 1.5 depicts a noncontiguous cartogram (Olson 1976). Polygons are moved together and changed in shape and relative position on contiguous cartogram maps.

Variations of these same symbol types are used on general reference maps, but the major emphasis is on qualitative (e.g., land, water) not quantitative differences between features. For example, a point symbol is used to represent the location of a city, a line symbol to represent a river, and a shading or color to represent a state or country. However, general reference maps may also depict a type of quantitative information. A road map, for example, will make distinctions between small, medium, and large roads by varying the thickness of the line. The size of cities may be similarly indicated with small, medium, or large dots.

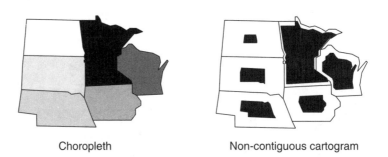

Choropleth   Non-contiguous cartogram

Figure 1.5   Examples of area symbols. The choropleth method is a common form of mapping. Shadings are used to depict value by area. The method implies that the value is constant throughout the area, which is almost always not the case. The less common noncontiguous cartogram varies the size of the area to depict value.

1.4.7 Classification of Data

The classification of data, a form of generalization on maps, is often necessary depending upon the method of symbolization. With the choropleth map, for example, only a limited number of shadings are usually available. For this reason the data is usually grouped into categories with one of a variety of classification methods (for a review of classification methods, see Dent 1993).

The computer has made it possible to create symbols of almost any magnitude, thereby obviating the need for data classification. Graduated point, line, and area symbol maps no longer require classification. The computer even makes it possible to create a continuum of shadings for the creation of unclassed choropleth maps (Tobler 1973, see Chapter 5).

The necessity for the classification of symbols on thematic maps is still a topic of controversy in cartography. Some believe that it is important for a map to create a more generalized representation and therefore argue for the necessity of classification. Others believe that a more accurate depiction of the data is the primary consideration and discount the importance of data classification. It may be that classification of quantitative data is merely an artifact of the hand-drawn, printed map. Classification of data may be an example of how the medium of paper has influenced the message of maps.

1.5 THE MEDIUM IS THE MESSAGE

There has been some rethinking recently about what computers represent to us. Hightech visionary Alan Kay, who conceived of the Dynabook and whose design work led to the development of the graphical user interface, argues that the computer is not a tool or an instrument but a medium. He cites Marshall McLuhan's contention from the early 1960s that electronic technology is not merely another form of communication but is the medium of our time and is reshaping and restructuring all aspects of our life.

A medium is the carrier of information. It is used to transmit knowledge and ideas between people. McLuhan's main concern is with the pervasive effect of the medium. He is particularly critical of the written word because it has forced us to attend to the recognition of text at the expense of all other sensory stimuli. This sensory impoverishment brought about by writing was further magnified by printing. McLuhan argues that the linear regularity of the printed page and our long-standing exposure to such a display has trained us to accept ideas only insofar as they conform to certain strict patterns. Thus, we have the creation of Gutenberg Man, a reference, of course, to the Gutenberg Bible and the invention of printing. Gutenberg Man, by McLuhan's account, is explicit, logical, and literal; by allowing himself to become overdisciplined by the closely ranked regiments of text, he has closed his mind to the wider possibilities of imaginative expression (Miller 1971, 5).

McLuhan states that literate people are visually incompetent because they have been so conditioned by the recognition of printed text. To demonstrate the extent to which we are conditioned by text, write the words *yellow, red, green,* and *blue* on a piece of paper with different colored pens. Write the word *yellow* with red, *red* with green, *green* with blue, and *blue* with a yellow colored pen. When you are finished, point to each word individually and say the color that was used to write the word. Notice how you could not help but read the word. The word *yellow*, for example, is written with red ink and you are supposed to say "red" when you look at the word. Because we are so conditioned by text, there is a tendency to read the word rather than say the color. This phenomenon has been studied in psychology and is called the Stroop Effect (Stroop 1935).

To demonstrate our lack of visual aptitude for things other than text, McLuhan cites studies of illiterate people that show a very high degree of visual competence within the area of prescribed social interests. The TEWA Native Americans of New Mexico, for example, have distinct names for a very large variety of coniferous trees. So many, in fact, that it is beyond the capability of literate people to see the differences. This visual ineptness of Western society may also be a reason that many avoid using maps.

According to McLuhan, some methods of communication are more pervasive than others, depending on the degree to which the medium employed reproduces the full sensory variety of the original experience. The capacity of any medium to perform in this way depends upon the environment it creates through the number of sensory channels that are called into play (Miller 1971, 3). The larger the number of senses involved, the more conducive the environment to convey a truer message. McLuhan states that the ratio of the senses is altered by each technology. All media alter the sensory mix and result in forcing changes on the individual (McLuhan 1967, 30). "Media, by altering the environment, evoke in us unique ratios of sense perceptions. The extension of any one sense above the others alters the way we think and act — the way we perceive the world" (McLuhan 1967, 41).

Further, McLuhan argues that we live in a rear-view mirror society (Theall 1971). He states that all new forms of media take their initial content from what preceded them. Not only is the new medium based upon the old, but society dictates that the only acceptable way of approaching the new medium is by emulating the old — through the rear-view mirror.

McLuhan has forced us to recognize the way in which technical innovation like printing and the electronic media create psychological environments, environments to which we subordinate ourselves without clearly recognizing the price we pay in doing so (Miller 1971, 8). Thus his famous motto — the medium is the message (McLuhan 1967). While it is an exaggeration, of course, to claim that the medium is the message, the medium does exert an effect over and above that which is carried in the message itself. According to McLuhan we have subordinated ourselves to printed text. Electronic communication helps to free us from the constraints of this medium.

1.5.1 The Medium in the Map

To what extent do maps suffer from the same limitations as text? Have printed maps closed our minds to the wider possibilities of imaginative expression and communication about the spatial world?

In the many thousands of years since the first map was created and especially in the 500 or so years since the first map was printed, cartography has developed within the limitations and possibilities of the paper medium. The way that we have learned to depict the world is to a large degree a product of this medium. Different methods of symbolization were introduced to improve the representational qualities of maps on paper. General map design principles were incorporated to improve the user interface of maps to the maximum extent possible with this static medium. An implicit goal of map design was to improve the aesthetic or graphic quality of maps to maintain the interest of the map user.

Each map embodies paper in the sense that the cartographer has been controlled by centuries of experience with this medium. The influence of paper upon cartography is so pervasive that it is difficult to conceive of a map form that is not influenced by this medium, even those displayed on the computer. We might speak of a "paper thinking" that still dominates our view of maps — influencing how they are created and how they are used. It will take many years to transform cartography to the new medium before this "paper thinking" can be overcome.

In addition, many cartographers now realize that considerable effort is required to retrieve information from a printed map (Muehrcke 1990). This is essentially true with any type of map use. One form of map use relies on measuring devices to determine distance or area — a process called cartometry (Maling 1989). A more common type of map use depends on a mental pattern recognition to determine relationships and associations, a process facilitated in whole or in part by information previously acquired and stored in the form of a mental map. In both cases a great deal of effort is required by the map user while the map remains a passive object.

The transition to the interactive map is made possible by the computer. Computers can be used to improve the general human interface with maps. With an appropriate user interface the interactive map can provide a more effective and more meaningful access to information. One of the major problems in increasing map use with computers is access to computers. The concept of the Superboard can be used to envision a possible future medium for cartography.

1.5.2 Envisioning a New Medium

An obvious trend in the development of computers since the early 1950s has been an increase in performance and a reduction in size. Initially filling entire rooms, computers with greater speed and capacity now fit in the palm of a hand. In the early 1970s, Kay (1977) at the Xerox Palo Alto Research Center envisioned a handheld computer that he

Figure 1.6   Apple Computer's Newton Personal Data Assistant (PDA). The handheld computer recognizes handwriting, converts sketches to lines and geometric shapes, and acts as a communication interface to other computers and to E-mail. (Photo courtesy of Apple, Inc.)

called the Dynabook. Twenty years later that dream is a reality. Apple's Newton computer is the size of a paperback book and weighs about a pound (Figure 1.6).

A number of other companies have also introduced palm-top computers. All of these computers are characterized by their small size, pen or stylus input rather than keyboard, character recognition, and screen resolutions of about 100 dots per inch. Most of these models use one of two operating systems: PenWindows from Microsoft or Go Corporation's PenPoint, although the Newton uses Apple's Newt/OS operating system. They are all based on a stylus interface and are designed to interpret handwriting. For the processing of commands the operating system incorporates the recognition of meaningful gestures such as caret for insert, cross-out for delete, square-bracket for block-marking and a question mark for help.

The stylus interface represents a major advance in the interface with computers. In many ways the stylus interface differs more from the popular mouse interface than the mouse differs from the keyboard. As advanced as these new computers are, they are only the precursors of a whole new generation of computers. What might a future stylus computer look like?

1.5.3 Specifications of the Superboard

The coming years will certainly see great advances in what are called palm-top computers. As the size of the display area is increased to the size of an atlas, a truly portable computer of this type will take on greater significance in cartography. For purposes of discussion let us refer to such an atlas-sized computer as the Superboard.

The Superboard is within the current realm of possibilities, although it would still be too expensive to produce. It would be flat and portable, with a screen area of 42 cm by 30 cm (16.5 inches x 11.8 inches; Figure 1.7). The initial implementations may have only a black and white display because of video memory considerations. Current screen resolutions are about 100 dots per inch (dpi). The Superboard would have a display

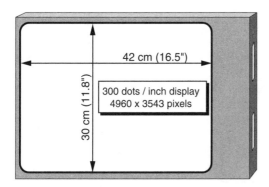

42 cm (16.5")

30 cm (11.8")

300 dots / inch display
4960 x 3543 pixels

Figure 1.7   Superboard computer. The proposed hand-held computer would have a large display area and a high-resolution screen.

resolution of at least 300 dpi, although 2000 dpi may be possible. The screen itself would be touch sensitive but a stylus pointing device or mouse could be attached for more precise command or drawing tasks. A graphic keyboard could be implemented on the screen to allow typing but the computer would normally not be used with a keyboard.

The initial stylus computers will be designed primarily for text applications and will have a relatively small screen. The Superboard version described here will require a number of years of development after the initial text-entry versions become successful. However, stylus computers may not become successful until they can support more sophisticated computer graphics applications.

To cartography the Superboard represents the medium beyond the microcomputer. The challenge cartographers will face is to use this new medium to its potential and not merely emulate paper methods, or even microcomputer methods, of map presentation. An example of its potential would be the implementation of a "cartographic zoom." Here the information content of the map would change as the scale is changed. With the type of zoom depicted in Figure 1.8 more features are added, and the lines become more detailed on the zoomed-in, larger-scale map. Icons of a magnifying glass with plus and minus symbols could be selected to control the zoom-in and zoom-out functions.

The Superboard could also be used to select interactively the features that are

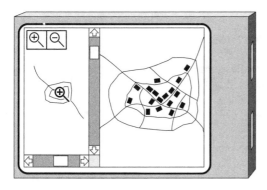

Figure 1.8   A "Cartographic Zoom." More features are added to the map by zooming on the smaller-scale display on the left. The magnifying glass icons indicate the direction of zoom.

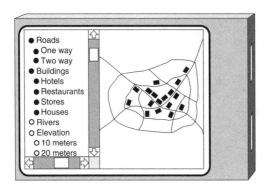

Figure 1.9   Selection of cartographic features. Features to be displayed on the map are selected from the list on the left. Selected features are indicated with solid dots.

displayed on a map. For example, a series of buttons could be used to control which features are visible (Figure 1.9). Clicking on a button would display the indicated features. This method of map display would greatly affect map design. It would reduce the amount of generalization of individual features, since not all categories would be displayed at once, as is necessary with maps on paper.

The stylus computer presents cartographers with a special challenge. As Muehrcke (1990, 13) states: "We must be willing to challenge all design assumptions associated with printed maps if we are to optimize the design of the new interactive map form." This will be especially true if stylus computers are to be effectively incorporated for map use.

1.6 SUMMARY

Each medium embodies certain possibilities and limitations for communication. In the many thousands of years since the first map was created, and especially in the 500 or so years since the first map was printed, cartography has developed within the limitations and possibilities of the paper medium. The way that we have learned to depict the world is to a large degree a product of this medium. A "paper thinking" still dominates our view of maps and controls the way maps are made and used.

The role of the computer in cartography is changing. Once used exclusively as a tool to help the cartographer make maps on paper, the computer is becoming a medium of communication. Maps are being viewed directly on the screen of the computer. This new medium has fundamental implications for how we portray the world with maps.

McLuhan, whose famous motto — the medium is the message — has come to symbolize the effect that the medium can have on communication, observes that we initially approach a new medium by emulating the old. The challenge presented to us by a new medium is to exploit its potential. The main advantage of the computer medium is interaction — the ability to choose what is displayed and how. A further advantage of the computer medium is its ability to create and display animations that transcend the static nature of the paper map.

The paper map will likely remain a major method of communicating information about the spatial world. Paper maps are easy to carry, they can be folded, and the cost of duplication is minimal. In the coming years the advantage of portability and cost of the paper map may become less important. In the meantime we should not let the advantages of portability and cost deter us from searching for better ways of communicating with maps or hinder our use of other media with which to do so.

1.7 EXERCISES

1. It is estimated that over 60 percent of the population do not use maps. Would you characterize yourself as a user of maps? Suppose you are driving in an unfamiliar city and become lost. You have a map of the city in the car. Would you attempt to use the map or would you stop to ask for directions? Why?

2. Why do you think that people have difficulties using maps? Can problems in map use be resolved through better education, or are there intrinsic problems with maps on paper?

3. How has the printed medium influenced the way maps are created and the way they are used? How has communication about the spatial world been limited as a result?

4. Are we conditioned by the printed map to such an extent that we cannot conceive of other, perhaps more informative, ways of representing the spatial environment? Can you image any innovative and more informative ways of representing the spatial environment?

1.8 REFERENCES

BAGROW, L., AND SKELTON, R. A. (1964) *History of Cartography.* London: Watts.

DENT, B. (1993) *Cartography: Thematic Map Design.* Dubuque, IA: Wm. C. Brown, 1993.

DOWNS, R. M. AND STEA, D. (eds). (1973) *Image and Environment.* Chicago: Aldine.

HARLEY, B. (1988) "Silences and Secrecy: The Hidden Agenda of Cartography in Early Modern Europe, *Imago Mundi,* vol. 40: 57–76.

KAY, A., AND GOLDBERG, A. (1977) "Personal Dynamic Media." *Computer* (USA): 31–41.

MACEACHREN, A. M. AND MONMONIER, M. S. (1992) "Introduction." *Cartography and Geographic Information Systems* 19, no. 4: 197–200.

MALING, D. H. (1989) *Measurements from Maps: Principles and Methods of Cartography.* Oxford: Pergamon Press.

McLUHAN, M. (1967) *The Medium is the Massage.* New York: Bantam.

MILLER, J. (1971) *Marshall McLuhan.* New York: Viking.

MORRISON, J. L. (1978) "Towards a Functional Definition of the Science of Cartography with Emphasis on Map Reading." *The American Cartographer* 5: 97–110.

MUERHCKE, P. C. (1990) "Cartography and Geographic Information Systems." *Cartography and Geographic Information Systems* 17, no. 1: 7–15.

OLSON, J. M. (1976) "Noncontiguous Area Cartograms." *The Professional Geographer* 28: 371-380.

PETCHENIK, B. B. (1979) "From Place to Space: The Psychological Achievement of Thematic Mapping." *The American Cartographer* 6, no. 1: 5–12.

ROBINSON, A. H. (1982) *Early Thematic Mapping in the History of Cartography.* Chicago: The University of Chicago Press.

ROBINSON, A. H. (1952) *The Look of Maps.* Madison, WI: The University of Wisconsin Press.

STROOP, J. R. (1935) "Studies of Interference in Serial Verbal Reactions." *Journal of Experimental Psychology* 8: 643–662.

THEALL, D. F. (1971) *The Medium is the Rear View Mirror.* Montreal: McGill-Queen's University Press.

TOBLER, W. (1973) "Choropleth Maps without Class Intervals?" *Geographical Analysis* 3: 262–265.

TUAN, Y. F. (1974) *Topophilia: A Study of Environmental Perception, Attitudes, and Values.* Englewood Cliffs, NJ: Prentice Hall.

WOOD, D. (1992) *The Power of Maps.* New York: Guilford.

WOOD, M. (1972) "Human Factors in Cartographic Communication." *Cartographic Journal* 9: 123–132.

FURTHER READINGS:

MONMONIER, M. S. (1993) *Mapping It Out.* Chicago: University of Chicago Press.

MONMONIER, M. S. (1991) *How to Lie with Maps.* Chicago: University of Chicago Press.

MONMONIER, M. S. (1985) *Technological Transition in Cartography.* Madison: University of Wisconsin Press.

MUEHRCKE, P. C. AND MUEHRCKE, J. O. (1992) *Map Use: Reading, Analysis and Interpretation*, 3rd. ed., Madison, WI: JP Publications.

ROBINSON, A. H. AND PETCHENIK, B. P. (1976) *The Nature of Maps: Essays toward Understanding Maps and Mapping.* Chicago: University of Chicago Press.

2

The Human Geographic Information System

2.1 INTRODUCTION

Every individual processes geographic information. The very act of walking or driving necessitates the input, processing, and display of information about the space around us. Our ability to function in space, and particularly our capacity to use maps to conceive and process spatial information beyond our own direct experience, allows each individual to be thought of as a human geographic information system (HGIS).

As we develop more interactive and animated methods of map display, it is particularly important that we understand how maps communicate. Geographic information systems serve as a valuable model in understanding the human processing of information from maps. In examining the similarities and differences between the human and computer GIS, we will find that the two share a number of functional and structural similarities. Theories from cognitive psychology concerning visual information processing and mental imagery help us to understand the human system. An examination of the visual processing of motion is important to understanding the interpretation of animated displays.

2.2 THE GIS WITHIN

Automated geographic information systems may be viewed as the external counterpart of an internal information system. Maps serve as a primary source of information for both the computer and human GIS. To examine the similarities more closely, it is useful to first review the computer system.

2.2.1 Computer Geographic Information Systems

Geographic information systems have developed since the early 1960s in association with the computer handling of geographic data. An early example was the Canadian Geographic Information System begun in 1963. Numerous systems have developed since the early 1980s with the availability of more functional software packages. The development of these systems has since turned into a multibillion dollar business, having a strong influence upon the disciplines of geography and cartography alike.

The general function of an automated GIS is to acquire, store, manipulate, and display geographic data for decision-making (Calkins and Tomlinson 1977). Maps constitute a primary source material for such systems, and a great deal of effort is involved in converting maps to a digital format for storage and processing by such systems. This conversion is usually accomplished by a conceptually important splitting of information content into "image data" and "descriptor data" components (Figure 2.1).

The input of data involves two distinct processes: (1) the image data set is created by digitizing maps with a digitizer or a scanner, thereby converting the graphic data to x and y coordinates or grid format (see Chapter 4), and (2) the descriptor data consists of textual information such as the classification of areas, area measurements, and feature names that are entered through a keyboard. Data is stored in a data base consisting of image and descriptor files. Standard data base procedures are used for storing and

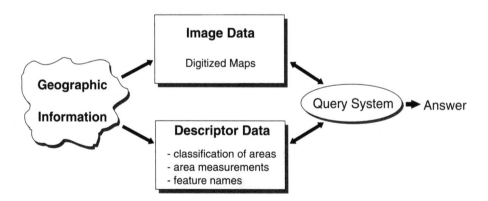

Figure 2.1   Simplified model of a geographic information system. Geographic information is split into image and descriptor data. A query system consults these data stores to answer questions.

processing the descriptor files. The manipulation of the image data is augmented by specific procedures for the handling of spatial data. Information is requested through a collection of routines that query the system, accessing the image and descriptor data sets in varying degrees depending upon the type of query.

2.2.2 The Functional Similarities

Humans may be compared to a computer GIS in two ways. In the first sense the ability to move through the environment uses all of the functions typical of a GIS. A person's every movement is dependent upon the input, manipulation, and display of spatially coded information. The input of "data" from the real world, the processing and manipulation of this data, and ultimately its display as a mental map are all functions attributed to a GIS.

The second type of GIS within humans is the ability to use maps to understand the world beyond our own direct experience. The ability to read maps and store some of the information in the form of a mental map is a type of data input. Results of experiments in cognitive psychology strongly indicate that such information is also split into image and descriptor (called *propositional*) components and stored as separate entities.

2.2.3 The Structural Similarities

The human and computer GIS not only share functional properties, they are characterized by structural similarities as well. Both the computer and human GIS have hardware and software components. Every computer GIS consists of at least an input device (keyboard, digitizer), a processing unit, and a disk storage device. The analogous structural features of the human system would be the eye, mental processes, and memory. In both cases the hardware may be viewed as the "fixed components" of the system.

Another structural similarity is the basic distinction between image and non-image components. The human mind seems to process both image and non-image information. The precise coding of the data is certainly different. It is unlikely, for example, that the human encodes image data numerically as is done with the computer. But both systems are able to reconstruct an image from the underlying codes — whatever the "deep representation."

The capabilities of a computer GIS is a function of its software. It is the programs that make the input, processing, and display possible. The software of the human system can be referred to as cognition. The term *cognition* is an all-encompassing concept that is generally defined as the "intelligent processes and products of the human mind" (Flavell 1977). Cognition includes such mental activities as perception, thought, reasoning, problem solving, and mental imagery. Neisser (1967, 5) defines cognition as "all the processes by which a sensory input is transformed, reduced, elaborated, stored, recovered and used." Improvements in computer GIS are made by improving the software. A more functional human GIS could result through greater training and use of the associated cognitive processes.

2.2.4 GIS and HGIS Differences

While certain functional and structural similarities can be identified, the capabilities between the two systems differ significantly. In creating a computer GIS one is concerned with accuracy and the ability to perform statistical analysis. The human system is satisfied with more general information. The difference is related to the purpose of each system. The computer system is perhaps a result of shortcomings in the human to deal with more detailed information.

One of the most striking differences between the human and its computer counterpart is the relative passivity of the computer system. In creating the computer GIS, data must be carefully selected and entered into the computer and then stored in separate files. Maps are input through a time-consuming digitizing process. In contrast, the human quickly selects the important features and internalizes the map as an image. Most importantly, the human does not store the information as separate entities but makes instant associations between newly entered and previously acquired information. Creating associations may be the only way that information can be remembered. Compared to the computer, the human is very active in acquiring spatial information and particularly adept at creating associations.

The active nature of human spatial information processing is apparent from evidence suggesting that locational information is coded into long-term memory without attention or conscious awareness (Mandler, et al. 1979). For example, we might remember the location of an article in a newspaper without intending to do so or the location in a notebook of an answer to a question without remembering the answer. Although it is possible that we may be consciously aware of this sort of spatial information, it is unlikely that we intend to commit it to memory. It is probable that the human system unknowingly encodes a great deal of locational information from maps. There is no comparable "effortless" mechanism in the computer system.

2.3 VISUAL INFORMATION PROCESSING

Compared to the computer, humans are particularly adept at processing visual or spatial information. To better understand the human geographic information system, we need to examine how visual information is processed. The processing of visual information, particularly the recognition of patterns, is not completely understood. Cognitive psychologists have presented a number of theories.

2.3.1 Stage Model

The stage model of human information processing envisions a series of memory stores, each characterized by a limited amount of information processing (Klatzky 1975). Like all models, this "stage model" is simplistic, but it serves in identifying distinct and largely verifiable memory stores in information processing. The three memory stores in this model are referred to as: (1) the sensory register, (2) short-term memory (STM), and (3) long-term memory (LTM). For visual information processing these three memory stores are referred to more specifically as iconic memory, the short-term visual store (STVS) and long-term visual memory (LTVM).

Figure 2.2 depicts these information processing stores in the recognition of a state boundary. According to this model, human information processing of visual information begins with iconic memory that is thought to hold information in sensory form for about 500 milliseconds (Humphrey and Bruce 1989, 196), long enough for it to be initially recognized. Iconic memory is a type of physical image within the retina that is of relatively unlimited capacity and is unaffected by pattern complexity. The short-term visual store is of much longer duration but is of limited capacity, and is therefore affected by complexity.

Moving information from iconic memory to the short-term visual store requires attention. This process simply refers to the human capability to "tune-in" certain sources of information and reject all others. An example of the process would be the visual identification of an animal that is partially obscured by vegetation. Attention makes it possible to "mentally focus" on the animal and reject the surrounding stimuli. This focusing is based on information in long-term memory. Kosslyn and Koenig (1992, 57)

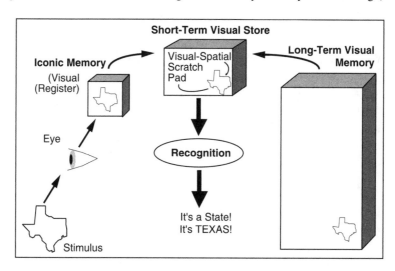

Figure 2.2   The recognition of a state outline. The processing of visual information begins with iconic memory where information is held for about 500 milliseconds. Visual information that has been recognized with the help of long-term visual memory is sent to the short-term visual store.

point out that stored information is used to make a guess about what we are seeing, and this guess then controls the attention process.

Visual information that has been recognized may be sent into the short-term visual store. Information is held here by a process called rehearsal that serves both to recycle material in STVS so that information does not decay and to transfer information about rehearsed items into long-term visual memory. Just as an "articulatory loop" is thought to provide additional storage for verbal material (the continuous internal repetition of a phone number, for example), a visual-spatial scratch pad (VSSP) is proposed as the equivalent "shelf" on which visual information is temporarily stored within STVS (Baddeley and Hitch 1974). Kosslyn and Koenig (1992) use the term *visual buffer* to describe short-term visual memory and point out that it can be activated from iconic memory or long-term memory. Whatever the source of information in STVS, it requires an active rehearsal for its maintenance and plays an important role in active visualization and imagery.

A number of theories exist concerning long-term visual memory. One theory suggests that LTVM is essentially a permanent storehouse — nothing is ever lost. Forgetting, it is argued, is a retrieval problem in which the appropriate link to a piece of knowledge is disrupted. Studies have shown an apparently limitless capacity to remember pictures. For example, Shepard (1967) showed subjects several hundred pictures from varied sources. Later, subjects were presented with a series of pictures and asked which they had seen before. A 98 percent accuracy rate was discovered. Another theory about LTM is that information is stored in a form compatible with sensory perception to facilitate the complicated process of pattern recognition.

2.3.2 Pattern Recognition

Visual information processing is based on the recognition of patterns. Visual pattern recognition converts the contents of iconic memory into something more meaningful through a matching process with previously acquired knowledge stored in long-term memory. There are at least three models of visual pattern recognition:

> 1. Template matching. With this model the image in iconic memory is matched to long-term memory representations, and the one with the closest match indicates the pattern that is present (Humphreys and Bruce 1989, 61).

> 2. Feature detection. This model specifies mini-templates or "feature detectors" for simple geometric features. An example of this type of model is the Pandemonium model proposed by Selfridge (1959; see also Lindsay and Norman 1976), which incorporates a series of "demons" or intelligent agents that work on a pattern by breaking it down into subcomponent features (Figure 2.3).

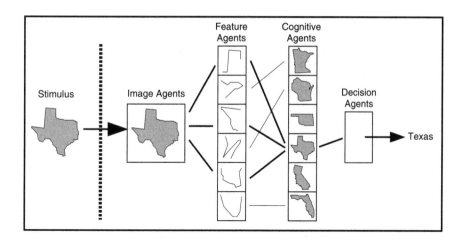

Figure 2.3   The Pandemonium model for recognizing the outline of Texas. The model proposes that shapes are characterized by significant features. Feature agents search for these features. Cognitive agents associate the features with the complete shapes and decision agents make the match and attach a name.

3. Symbolic description. A third model proposes that objects are represented symbolically as structural descriptions. A structural description can be thought of as a list of propositions or facts that describe properties of the individual parts of an object and their spatial relationships. For example, a propositional model for the letter T is "a horizontal line that is supported by a vertical line, and that this support occurs about half way along the horizontal line" (Bruce and Green 1990, 186). However, this model does not clearly specify how a pattern is recognized based solely on its structural description (see Pinker 1985). Also, it is difficult to see how a propositional model could work for complex shapes like the outline of a state.

2.3.3 Computational Model

A slightly different way of thinking about image interpretation is given by Marr (1982). Marr's theory incorporates a computational approach based on neuron like models of visual perception. These models have the advantage of being biologically plausible, and provide a mechanism to incorporate parallel computations in visual information processing. Marr attempts to explain visual information processing in a framework that cuts across the boundaries of physiology, psychology, and artificial intelligence.

According to Marr very general knowledge about the physical world is used to interpret the retinal image. Perception, he argues, is based on assumptions that are generally true of the world. These assumptions include the fact that surfaces are generally smooth or that light comes from above — assumptions that a visual system would not have to discover from experience but for which it is "hard-wired" (i.e., innately specified).

An important part of Marr's theory is that a number of different representations must be constructed from the information in the retinal image. The first representational stage, called the *primal sketch*, captures the two-dimensional structure of the retinal image. The next representation, called the *2 1/2 D sketch,* describes how surfaces are oriented with respect to a viewer. Finally, Marr argues that a representation of the three-dimensional shape of the object must be constructed. This last representation is called the *3D model.*

Marr and Nishihara (1978) present a theory of object recognition based upon this general theory. They argue that objects or their components can be constructed from generalized cones or cylinders and that the lengths and arrangements of the axes of these components relative to the major axis of the object as a whole can be used in object recognition. They present the example of discriminating between a human and a gorilla. If each is viewed as consisting of generalized cones that are used to represent the head, trunk, arms, and legs, the two can be differentiated based on the relative lengths of the axis of these components.

Marr and Nishihara go on to suggest that a configuration of cones and cylinders derived from an image could be matched against a catalog of different object models distinguished on the basis of the number and disposition of their component parts. A major advantage of this approach is that descriptions at different spatial scales can be constructed. Humphreys and Bruce (1989, 72) point out that each entry in the catalog could point to a hierarchically organized set of 3D models at different scales and that the catalog could be accessed at any one of these levels (Figure 2.4).

Marr's theory of visual information processing and object recognition is based on

Figure 2.4   A hierarchy of 3D models with increasing levels of detail. Objects are represented with different levels of generalization.   (After Humphreys and Bruce 1989, 71.)

breaking down the retinal image into its component parts. These parts are not propositional descriptions but simpler shapes. The process is analogous to generalization in cartography. Maps are based on a simplification of reality. It seems that the recognition of patterns is based on much the same process. This simplification process has been confirmed in studies on the mental image.

2.4 THE MENTAL IMAGE

In psychology the mental image is defined as an internal representation similar to that resulting from sensory experience but arising from memory. Although we may speak of such things as auditory mental images, olfactory mental images, it is the "visual mental image" that is referred to here, and it is this form of mental image that has been the focus of most psychological investigations.

Methodological advances have enabled researchers to examine mental events in more objective ways. As a result, new empirical findings about mental imagery have emerged. The purpose of research in mental imagery has been to (1) determine the existence of mental images, (2) define their properties, and (3) establish how they are used in thinking.

2.4.1 Existence of Visual Codes in Memory

A number of studies have established the existence of visual codes in memory. In an often cited study Shepard and Metzler (1971) measured the time it took for subjects to decide whether two complex block structures depicted the same three-dimensional shape (Figure 2.5). It was found that the amount of time it took to make the judgments increased linearly with the difference in rotation between the two objects. The rate of rotation was found to be approximately 60 degrees per second. The researchers

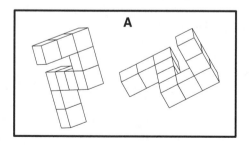 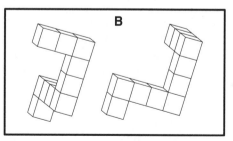

Figure 2.5   Block structures used in Shepard and Metzler's object rotation experiments.   It was found that the amount of time required to make a judgment of similarity or difference increased linearly with the difference in rotation between the two objects. The figures in example A are identical but are rotated by 80 degrees. Figures in example B are not the same. (After Shepard and Metzler 1971.)

concluded that the subjects were rotating the visual image of one figure into congruence with the other and that the farther the objects had to be mentally rotated, the more time was required for the decision. Like physical rotation, mental rotation takes longer the greater the distance involved, suggesting an internal imagery space with properties analogous to physical space.

Another line of research concerning the existence of visual codes in memory suggests a difference between visual and linguistic memory. Paivio (1971) first postulated that visual imagery and verbal symbolic processes represent separate coding systems in memory. Moreover, visual representations are not restricted to material that is intuitively spatial, such as a map; rather there is a distinction between material that is concrete and readily imageable and material that is abstract and cannot be readily associated with an image (McTear 1988, 106). Paivio (1986) showed that the image-evoking value of nouns, referred to as "imagery concreteness" (e.g., "apple" leads to a more concrete mental image than the more abstract "distance"), correlates highly with the ability to remember. It appears that words coded under both image and verbal mediators will be more easily remembered than words that are only verbally coded since more "links" exist to those words in memory.

It would therefore be useful to consider images as a distinct form of representation in terms of their functional role in cognition. What remains to be examined are the specific properties ascribed to mental images.

2.4.2 Properties of Mental Images

A variety of studies in psychology have attempted to define the properties of mental images. Researchers have theorized that imagery uses representations and processes that are ordinarily associated with visual perception. Kosslyn (1980) theorizes, in particular, that mental images are analogous to displays on a graphic computer terminal or cathode-ray tube (CRT) on which graphic displays can be generated. In his view the mental image is a data structure consisting of two parts: (1) the "surface representation" or quasi-pictorial entity that we experience and (2) the "deep representation," the information in memory from which the surface image is derived. It follows from the CRT metaphor that mental images have two important structural components: (1) they have analog properties, that is, they preserve relative interval distances between their component parts and 2) they have a limited size.

In a series of experiments Kosslyn demonstrates that mental images are characterized by these two structural components. In experiments involving visual scanning, the time required to scan between objects in mental images is used to determine whether mental images have true analog properties. In one of these experiments a fictional map with seven locations (among them house, well, tree, lake) is presented to subjects (Figure 2.6). Subjects are repeatedly asked to draw the map on a blank piece of paper until objects were depicted to within 0.25 inch (0.635 cm) of their actual location. Next, subjects are asked to image the mental map and mentally "stare"

Figure 2.6   Map used in Kosslyn's mental scanning experiment. Subjects were asked to image a mental representation of this map and "stare" at a named location. Then a second name was presented and the reaction time measured for the subject to scan to the object. It was found that the time to scan to the named location increased at a constant amount with the distance to be scanned. (After Kosslyn 1980, 43)

at a named location. A second name is presented, and the reaction time is measured for the subject to scan to the object. It was found that the time to scan increased linearly with the distance to be scanned (Kosslyn 1980, 44; Tye 1991, 44). Kosslyn concluded that mental images, like maps, depict information about interval spatial extents.

If mental images have limited spatial extent as theorized, the mental image must contain a limited number of features (in the same sense that a map is limited in the number of features it can reasonably show). Kosslyn devised a procedure in which size was manipulated indirectly by asking subjects to image "target" animals, such as a rabbit, as if it were standing next to either an elephant or a fly (Figure 2.7). The assumption was that if a mental image takes up a fixed amount of space and most of this space is taken up by the elephant, the adjacent scaled image of the rabbit should be subjectively smaller and not as detailed. Accordingly, if a rabbit is scaled next to an image of a fly, the rabbit should seem much larger, with more defined features. The results of the experiment showed that people required more time to mentally "see" parts of the subjectively smaller animals, that is, the animals imaged next to the larger animals.

(a)   (b)

Figure 2.7   Animals used in Kosslyn's selection of features experiment. In (a) a rabbit and elephant are imaged. In (b) a rabbit and a fly are imaged together. It was found that more time was required to image features of the smaller object. (After Glaser, et al. 1979, 373.)

Subjects explained that they had to "zoom in" to see the properties of the subjectively smaller images. Kosslyn concluded that a selection of features had occurred in the originally smaller (before "zoomed") image because the mental image has a fixed size and resolution (Kosslyn 1983, 56).

2.4.3 Visual Cognition

Visual cognition is defined as the use of mental imagery in thinking. Mental images can be derived from material stored in memory, or they can be mental "drawings" of things we have never seen. Kosslyn and Koenig (1992, 129) state that imagery can be used in the following ways:

1. Reasoning. Reasoning with imagery is often reported by scientists or inventors. It is based on the creative generation of images. This may be accomplished by combining familiar elements in new ways or from scratch using only elementary components. The use of imagery in reasoning involves not only generating and inspecting imaged objects but transforming and retaining images long enough to work with them (Kosslyn and Koenig 1992, 146). An example might be the image that an electrician might form of a wiring network in a house.

2. Learning a skill. Learning a new skill can be aided through the use of imagery. In this case an image is used to define physical movements (Kosslyn and Koenig 1992, 155). The technique is often used in sports training. Shooting baskets, for example, can be improved by using imagery as a part of practice. This aspect of visual cognition may be used to understand the effect of a map in controlling our movements in space.

3. Comprehending verbal descriptions. A mental image can be used to interpret verbal descriptions. The description may consist of a floor plan to a house or instructions on how to travel to a certain place. A mental image seems to be vital in the interpretation of the description.

4. Creativity. Finke (1989) shows how images can stimulate the discovery of unexpected patterns, new inventions, and creative concepts. This form of visual cognition involves some type of discovery.

2.4.4 Maps and Mental Imagery

Mental images and maps seem to share unique properties that make them intellectually useful and distinctive from other forms of representation and communication. There are still remaining questions concerning the relationship between maps and mental images — how they are used and how they combine with linguistic symbolism in thought processes.

Results of experiments in cognitive psychology indicate a strong relationship between images and maps. The result of Kosslyn's first experiment shows that images, like maps, represent relative interval distances. More significantly, perhaps, are the results of the second experiment, which seem to indicate that mental images are not copies of sensory impressions, like "pictures in the head," but rather seem to be intellectually processed and generalized representations analogous to maps. If there is indeed a selection of features in the smaller, before-zoomed image of the animals, there would be a very strong relationship between mental images and maps. Mental images may be defined as the internal counterpart of maps in the sense that both are characterized by the selection of features.

The zooming of mental images, and particularly the addition of more features as mental images are made larger, is an interesting mental process. The fact that the human mind can zoom into the image relatively effortlessly would make a strong case for more interactive methods of map display that could simulate this relatively complicated process. It seems that the processing and display of mental images within the brain is far more dynamic than is possible with the printed map.

2.4.5 Mental Imagery and Linguistic Symbolism

The distinction between linguistic or propositional and graphic symbolism is central to cognitive psychology and leads to some interesting conceptions of how images are actually used. It has been suggested that the relative processing speed of the two kinds of representations may determine which is used to answer a question. As with the Selfridge model of pattern recognition, one can think of "demons" working in both domains, each attempting to answer a given question as quickly as possible. For example, in answering whether Volkswagen Beetles have tires, a propositional-linguistic file may be used, since VWs are cars and cars have tires. Consider what happens, however, when one decides whether VW Beetles have vent-wing windows (little triangular windows at the front of the door). It is unlikely that this information is explicitly noted in the propositional file associated with Volkswagen Beetles. Thus one may have to image the car in order to retrieve the information (Kosslyn and Shwartz 1977, 284).

Putting the example in geographic terms (Peterson 1987, 39), a question of whether Houston and San Antonio are both in the state of Texas may be most quickly answered by consulting a propositional file (Figure 2.8a). Confronted with a question of whether Houston is north of San Antonio, however, one may need to consult an image file (Figure 2.8b). This suggests that whenever "some previously unconsidered spatial relation (which hence is unlikely to be coded propositionally) is queried, and this relation is implicit in the encoded image" (Kosslyn 1980, 397), one uses a mental image to answer the question.

(a)
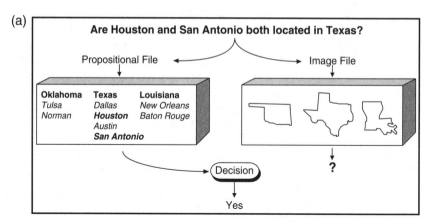

(b)
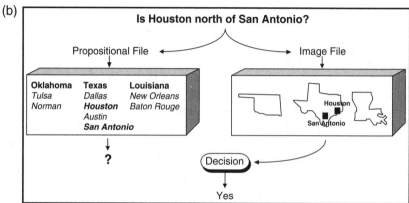

Figure 2.8 The role of linguistic and graphic symbolism in answering map-related questions. In (a) linguistic symbolism is used to answer the question of whether Houston and San Antonio are both in Texas. In (b) graphic symbolism is used to answer the question of whether Houston is north of San Antonio.

2.5 VISUAL MOTION PROCESSING

Cognitive psychologists have concentrated on the mental processing of static images. Of course, the world is dynamic and the mind must be able to process moving images. In addition, it has been noted that even when an observer views a scene containing only motionless objects, the observer's eyes will be in continuous motion. Heckenmuller (1965) found that if the effects of such eye movements are neutralized by artificially stabilizing the image on the retina, we become blinded. Motion, either of objects in the environment or created artificially by the movement of the eyes, is essential to perception. Psychologists have taken at least three separate approaches to the understanding of motion processing.

2.5.1 Apparent Motion

It has long been known that the perceptual system can be made to perceive motion when none actually exists. This principle is the basis of both film and displays on a computer terminal. Film projects 24 still pictures per second, and the standard video displays about 30 to 60 frames per second. In both cases the mind constructs the perception of motion from a series of static images.

The phenomenon provided a justification for analyzing the perception of motion by determining how correspondences are established between the "snapshots" of the changing retinal patterns that are captured at different instances. The major conclusion of these studies is that correspondences between individual elements in successive displays are established on the basis of the primitive elements of figures (primal sketch) rather than between whole figures (Julezs 1971; Marr and Poggio 1976). Ullman (1979) provides a computational account for how correspondences are made, using the principle of "minimal mapping." He defines an affinity measure between pairs of elements in a display. To solve the correspondence process in a display with several elements between frames, a solution is found that minimizes matches with poor affinities and maximizes those with strong affinities. When the affinities between elements are high and those elements change position between frames, we have the perception of motion.

2.5.2 Ecological Approach to Visual Perception

The importance of motion perception is one of the major influences in the development of Gibson's (1979) theory of ecological perception. He argues against the predominant view that visual perception involves the recognition of static retinal snapshots. Instead, Gibson claims that information to specify the structures in the world is detected "directly" from dynamic patterns of optical flow. In other words, perception is dependent on change in the display.

Gibson argues that it is the flow in the structure of the total optic array that provides information for perception, not the individual forms and shapes of the "retinal image." He emphasizes the perception of surfaces in the environment. Surfaces consist of textural elements, and different objects in the environment have different surfaces. The texture that exists in the surfaces structures the light that reaches the observer. Gibson argues that it is this structure in the light, rather than stimulation by light, that furnishes information for visual perception.

According to Gibson, the analysis of the movement of objects must begin with the analysis of the information available in the continuously changing optic array. To explain the phenomenon of apparent motion, Gibson argues that the stimuli preserve the transformational information that is present in the real moving display. He maintains that the stimulus information contains transformations over time that specify the path of the motion. Gibson's ideas have a great deal of applicability to the processing of dynamic displays and have influenced work on the mental processing of computer-generated displays (Wanger, et al. 1992).

2.5.3 Evidence from Physiology

It has long been theorized that the visual system has a special sensitivity to moving images. An early discovery about visual motion processing was made by Barlow and Levick (1965). They documented the existence of neurons in rabbits that are selectively sensitive to speed and direction of retinal image motion. It has since been generally verified that motion detection in the lower animals, particularly insects, is done by neurons in the retina. In primates, however, the perception of motion can be traced directly to higher levels of processing in the cortex (Movshon 1990, 126).

Recent physiological evidence points to special neural subsystems in the brain that handle the visual processing of movement. The first evidence of these subsystems was from people who had suffered bilateral damage to a certain part of the cortex. One patient, for example, had an impaired ability to see continuous movement (Zihl, et al. 1983). This person had difficulty crossing a road because she could not judge traffic speed. She had difficulty with conversations because she could not follow facial expressions. She also reported that when pouring tea into a cup, the tea appeared frozen "like a glacier." It was also difficult for her to know when to stop pouring because she could not detect the rising liquid in the cup. Despite these problems, she remained able to recognize stationary objects.

A physiological model of visual mental processing is analogous to the stage model of information processing, but relies on a more anatomical explanation. For example, the visual register corresponds to the eye, where receptors, stimulated by light, connect to ganglion cells that connect to individual fibers of the optic nerve called axons. The axons transmit their messages to the brain as electrical impulses. The impulses proceed to the primary visual cortex in the back of the brain after passing through a synapse, called the lateral geniculate nucleus, about half way along the journey to the cortex. Axons from the right half of the visual field of *both eyes* go to the left side of the brain, while those from the left fields go to the right half of the brain. These two sides of the visual field are reunited at the same place in the cerebral cortex.

The cerebral cortex plays an important role in the visual system. It has been found that in monkeys over half of this part of the brain is involved in the processing of visual information. This "visual" part of the cortex would seem to correspond to the short-term visual store. It is here that the images presented to the eye are actually "visualized." The cortex is made up of a large number of cells. In each eye, there are about 1 million ganglion cells in the human retina connecting to 1 million axons that go to the brain. These connect to about 100 million cells in the visual cortex (Barlow, et al. 1990, 12).

It seems that there are two main groups of retinal neurons feeding into separate pathways that can be traced directly to the cerebral cortex. One of these "visual pathways" appears to specialize in analyzing motion. The other seems to be more concerned with the form and color of visual images (Movshon 1990, 127). Evidence in support of a motion pathway was determined through the functional analysis of neural activity. The evidence seems to suggest that the visual processing of motion is accomplished separately from the processing of the attributes of objects, such as form and color.

2.6 SUMMARY

Communication with maps can be modeled in the form of a human geographic information system. It seems that much of the information derived from a map is in image form — sometimes encoded without conscious awareness — and may be used long after the map has ceased to be a visual stimulus. It seems probable that the mind is a storehouse of images from which previously unconsidered spatial information may be derived. The process of cartographic communication, therefore, continues in the absence of the map. The purpose of modeling cartographic communication in the form of a human geographic information system is to consider the long-term aspects of map use. Map reading is not an isolated, one-time activity. Much of the information derived from a map is used long after the map has been viewed.

In the cognitive framework perception and pattern recognition are closely linked with memory, and since pattern recognition occurs so quickly, many cognitive psychologists believe that the representations in memory have sensory qualities. Much evidence has accumulated in recent years for analog representations in the form of mental images. Kosslyn likens such images to displays on a CRT screen and has shown that mental images have at least two of the same spatial properties: (1) they preserve relative interval distance and (2) they have a limited spatial extent. The results of various experiments indicate that mental images are functional representations in human thought.

The importance of visual motion processing is just now beginning to be understood. It seems that parts of the visual cortex specialize in motion processing. That this is accomplished in the cortex and not in the eye, as with lower animals, indicates that training and experience are factors in our ability to detect motion. This would suggest that greater experience with dynamic displays could result in a greater ability to interpret these displays.

2.7 EXERCISES

1. The shapes on the right are states in the United States. Can you identify them? How many shapes like this can you identify, and how did you learn them?

2. Find the three errors in the outline of the continent of Africa. How are you able to identify these errors? Is your mental map of the outline of Africa more accurate than the map presented here?

3. On a separate piece of paper, draw the outline of the United States. Can you identify errors in your own map? How are you able to identify these errors? Why can't you draw a map of the United States that you feel is correct?

2.8 REFERENCES

BADDELEY, A. D. AND HITCH, G. (1974) "Working Memory." *The Psychology of Learning and Motivation*, vol. VIII, edited by G. Bower, New York: Academic Press: 47–89.

BARLOW, H., BLAKEMORE, C. AND WESTON-SMITH, M. (eds.). (1990) *Images and Understanding: Thoughts about Images, Ideas About Understanding.* Cambridge, England: Cambridge University Press.

BARLOW, H. B. AND LEVICK, W. R. (1965) "The Mechanism of Directionally Selective Units in the Rabbit's Retina." *Journal of Physiology* 178: 477–504.

BRUCE, V. AND GREEN, P. R. (1990) *Visual Perception: Physiology, Psychology and Ecology.* Hillsdale, NJ: Erlbaum.

CALKINS, H. W. AND TOMLINSON, R. F. (1977) *Geographic Information Systems, Methods and Equipment for Land Use Planning*, Washington: Department of the Interior.

FINKE, R. A. (1989) *Principles of Mental Imagery.* Cambridge, MA: MIT Press.

FLAVELL, J. H. (1977) *Cognitive Development*. Englewood Cliffs, NJ: Prentice Hall.

GIBSON, J. J. (1979) The Ecological Approach to Visual Perception. Boston: Houghton–Mifflin.

GLASER, A. L., Holyoak, K. J., and Santa, J. L. (1979) *Cognition*. Reading, MA: Addison-Wesley.

HECKENMULLER, E. G. (1965) "Stabilization of the Retinal Image: A Review of Method, Effects and Theory." *Psychological Bulletin* 63: 157–169.

HOCHBERG, J. (1986) "Representation of Motion and Space in Video and Cinematic Displays." In (Eds.) *Handbook of Perception and Human Performance* 2: 22–1 — 22–64, edited by K. R. Boff, L. Kaufman and J. P. Thomas, New York: John Wiley.

HUMPHREYS, G. W. AND BRUCE, V. (1989) *Visual Cognition: Computational, Experimental and Neuropsychological Perspectives.* Hillsdale, NJ: Erlbaum.

JULEZS, B. (1971). *Foundations of Cyclopean Perception.* Chicago: University of Chicago Press.

KLATZKY, R. L. (1975) *Human Memory: Structures and Processes.* San Francisco: Freeman.

KOSSLYN, S. M. (1983) *Ghosts in the Mind's Machine.* New York: Norton Press.

KOSSLYN, S. M. (1980) *Image and Mind.* Cambridge, MA.: Harvard University Press.

KOSSLYN, S. M. AND KOENIG, O. (1992) *Wet Mind: The New Cognitive Neuroscience.* New York: Free Press.

KOSSLYN, S. M. AND SHWARTZ, S. P. (1977) "A Simulation of Visual Imagery." *Cognitive Science* 1: 265–295.

LINDSAY, P.H. AND NORMAN, D.A. (1976). *Human Information Processing.* New York: Academic Press.

MARR, D. AND POGGIO, T. (1976). "Cooperative computation of stereo disparity." *Science* 194: 283-287.

McTEAR, M. (ed.) (1988) *Understanding Cognitive Psychology.* Chichester, England: Ellis Horwood.

MANDLER, J. M., Seegmuller, D. and Day, J. (1979) "On the Coding of Spatial Information." *Memory and Cognition* 5: 10–16.

MARR, D. (1982) *Vision.* San Francisco: Freeman.

MARR, D. AND NISHIHARA, H. K. (1978) "Representation and Recognition of the Spatial Organization of Three-Dimensional Shapes." *Proceedings of the Royal Society of London, Series B*, 200: 269–294.

MOVSHON, A. (1990) "Visual Processing of Moving Images." In *Images and Understanding: Thoughts About Images, Ideas About Understanding,* edited by H. Barlow, C. Blakemore, and M. Weston-Smith. Cambridge, England: Cambridge University Press.

NEISSER, U. (1967) *Cognitive Psychology.* New York: Appleton-Century-Crofts.

PAIVIO, A. (1986) *Mental Representations: A Dual Coding Approach.* Oxford: Oxford University Press.

PAIVIO, A. (1971) *Imagery and Verbal Processes.* New York: Holt, Rinehart and Winston.

PETERSON, M. P. (1987) "The Mental Image in Cartographic Communication." *The Cartographic Journal* 24: 35–41.

SELFRIDGE, O. G. (1959) "Pandemonium: A Paradigm for Learning." *The Mechanization of Thought Processes.* London: Her Majesty's. Stationery Office.

SHEPARD, R. N. (1967) "Recognition Memory for Words, Sentences and Pictures." *Journal of Verbal Learning and Behavior* 6: 156–163.

SHEPARD, R. N. AND METZLER, J. (1971) Mental Rotation of Three-Dimensional Objects. *Science* 171: 701–703.

TYE, M. (1991) *The Imagery Debate.* Cambridge, MA: MIT Press.

ULLMAN, S. (1979) *The Interpretation of Visual Motion.* Cambridge, MA: MIT Press.

WANGER, L. R., FERWADA, J. A. AND GREENBURG, D. P. (1992) "Perceiving Spatial Relationships in Computer-Generated Images." *IEEE Computer Graphics and Applications,* no. 5: 44–55.

ZIHL, J., VON CRAMON, D., AND MAI, N. (1983) "Selective Disturbance of Movement Vision after Bilateral Brain Damage." *Brain* 106: 313–340.

FURTHER READINGS:

HAMPSON, P. J. (ed.). (1990) *Imagery: Current Developments.* London and New York: Routledge.

LOGIE, R. H. AND DENIS, M. (eds.) (1991) *Mental Images and Human Cognition.* Amsterdam and New York: North Holland.

ROLLINS, M. (1989) *Mental Imagery: On the Limits of Cognitive Science.* New Haven: Yale University Press.

3

Maps for the Mind

3.1 INTRODUCTION

Consider for a moment how you answer the following questions: Do the states of New Hampshire and Vermont have similar shapes? What is the location of Atlanta within the state of Georgia? Which states border the Mississippi river? For people familiar with the United States, the mere mention of these places leads to the formation of a mental map. The human mind seems to be extremely active in acquiring, manipulating, and displaying mental images in the form of maps. Research in cognitive psychology has shown that these mental images are a functional part of human thought.

A zooming process may also accompany the experience of mental mapping. The mention of a city such as Los Angeles, for example, might result first in a mental map that locates Los Angeles within the United States, then within the state of California, and ultimately within the southern part of the state. This mental zooming indicates that the human mind can visualize not only static mental maps but also dynamic mental map *animations*.

Limited to the static medium of paper, cartographers have been unable to properly engage the mental imaging abilities of the human mind. A map form is needed that more closely approximates the dynamic nature of mental maps. Combined with an appropriate user interface, the computer represents such a dynamic medium. This chapter examines the possibilities of interactive and animated cartography using the computer.

3.2 INTERACTION AND THE USER INTERFACE

Interaction with computers was once simply defined by the amount of time required for the computer to respond to a command. It has since become apparent that *response time* is only one aspect of the human interaction with computers. The term *user interface* now describes the general interaction between humans and computers and includes the myriad factors that influence the acceptance and use of computers.

The word *face*, as in interface, is used in the sense of a dividing line or boundary. An example of a face would be the boundary between two different substances that do not mix like oil and water. If these two substances are placed in a container, a face or boundary forms between them. Interface refers then to the interactions across a face.

The general user interface with computers has evolved through switches, cards, keyboard, and finally to a variety of pointing devices, currently best exemplified by the mouse. The mouse is implemented as part of a graphical user interface (GUI) that consists of menus, windows, dialogs, and icons. All of these graphical user interface elements form the basis of the interactive computer.

The development of interactive computers is a relatively recent phenomenon. Throughout much of the 1950s, 1960s and most of the 1970s computer cards were the primary method of controlling the mainframe computer. An operator would enter the cards through a card reader, and the computer would generate a computer printout of the results. The entire process would require 15 to 30 minutes or more, depending primarily upon the speed of the human operator.

Early minicomputers and microcomputers were controlled through a command-line interface that required the typing of a specific sequence of characters through a keyboard. The keyboard was a vast improvement over computer cards, especially for those accustomed to working with computers. But for individuals who did not know the specific command lines the computer remained inaccessible.

It was the microcomputer and a pointing device that brought the technology of computers to a significant portion of the population. The Apple Macintosh, introduced in 1984, popularized a point-and-click interaction with computers. Programs for other operating systems that implement a graphical user interface, including Microsoft Windows for PC computers, are expanding the popularity of this form of human-computer interaction. The graphic input device has made interaction with maps possible. The computer can now respond to the clicks of a mouse to display a map.

The term *user interface* is most often used in reference to the interface between humans and computers but the concept of interface can be used to refer to all interactions that take place between humans and inanimate objects. A television set, for example, has a user interface in the form of switches, dials, and the buttons on the remote control unit. An effective interface for the television set includes symbols in the form of numbers and arrows to explain the operation of the buttons and dials. An interaction of sorts occurs between the user and the television as the set is used and the suspected operation of the switches and buttons is confirmed.

The word *interface* relates to maps in two ways: (1) the map is an interface to the world and (2) it is composed of user interface elements. The selection of colors, the legend, the map layout, its sectioning, and folding are all aspects of a map's user interface. Maps are evaluated on the perceived quality of their user interface. An interaction between map and user occurs as the map is used and its relationship to other maps or to the world itself is confirmed. For example, the use of a street map is based on the constant comparison between features on the map and the features in reality. If a map is well designed, the user interface incorporated in the design creates a map that serves as an effective interface to the world.

The relatively low level of user interaction is a problem with the printed map. The map user cannot change the view or choose what is presented. The printed map serves as a poor interface to the world.

3.3 THE INTERACTIVE MAP

The interactive map is a computer-assisted form of map presentation that attempts to mimic the display of mental maps in the mind. It goes beyond the mental display of maps by presenting a more vivid and accurate display. The maps include more features and do not exhibit the distortions and biases of mental maps. The interactive map is characterized by an intuitive user interface consisting of graphical icons, a pointing device and the near instantaneous display of maps. The interactive map includes "tools" to further zoom in on the map or "open-up" different areas and may include "video-clips" of places with pictures and sound. Ultimately, the interactive map is an extension of the human ability to visualize places and distributions.

3.3.1 Types of Map Interaction

A map consists of graphic (image) and attribute (descriptor) components. The representation of points, lines, and areas on maps is the graphic part of a map. Attributes define what is depicted and can be either qualitative (also called nominal and having distinct names) or quantitative (a data value such as population). The name of a road is a qualitative attribute, while its size, as represented by its width, would be quantitative. Map interaction is possible with both the graphic and attribute parts of a map.

Examples of interaction with the graphic component of a map would be a change in scale, a change in perspective, or a change in symbolization. A change in scale could be a simple zoom-in, zoom-out display. The change in perspective may involve an oblique view or the rotation around a three-dimensional object. An example of a change in symbolization would include changing the size or any other graphic characteristic of a symbol.

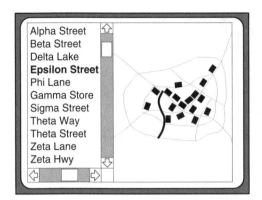

Figure 3.1   A possible interaction with attribute data. In this case street names are displayed in a separate window. Clicking on a street name highlights the corresponding street on the map. The interaction may also work in the opposite direction. Clicking on a street will activate the street name. By separating the text from the map the map becomes less cluttered, allowing other features to be depicted.

Incorporating the attribute data into the interactive map dramatically increases the number of manipulations that are possible. For example, a program could list street names and maps in two separate windows on the screen. Pointing to a street name from a list of names would highlight the street on the map and center the display (Figure 3.1). A simple on-off blinking of the street would identify its location. The reverse can also be accomplished — clicking on the street can serve to highlight the street name.

The attribute data can also be used to implement a *cartographic zoom*, as presented in Chapter 1. Increasing the scale of a map is usually accompanied by an increase in detail. More features are shown and the lines are less generalized. A normal graphic zoom does not add extra information; it simply enlarges the display. The attribute data could include a field that defines the scale at which the items should be depicted. Each point in the map could be similarly labeled. A great deal of cartographic research has examined the automation of line and feature generalization that would be necessary for this type of cartographic zoom (for a summary, see McMaster and Buttenfield 1990).

Interaction is also possible with quantitative attribute data. The first objective here is to create an association between the quantitative values and the map. For example, Figure 3.2 depicts two windows from a choropleth mapping program. A column of data is selected from the data window and then transformed into a map in the map window.

The second objective is to add interaction to the analysis of the data. This might involve changing the classification method or the number of classes. Figure 3.3 shows a graphical menu palette that is be used to select alternative classification methods or number of classes. Numerous types of data manipulation are possible with a single data set or between data, and these form the basis of interaction with quantitative attribute data (Figure 3.4).

Other types of interaction combine graphic and attribute data. For example, by adding a distance attribute to street segments the shortest distance between two locations can be calculated and then graphically presented. By specifying the speed of travel, an

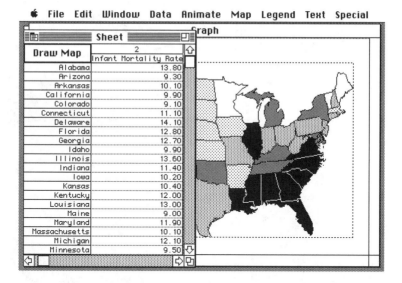

Figure 3.2   Creating an association between the map and the data. The window on the left lists the data values for each state. The shaded map was created from the classified data.

estimated arrival time can be determined and shown on the map. Finally, other media, including sound, pictures, and video can be integrated into the display of maps. Interaction can be incorporated into almost every aspect of map display, but its successful implementation requires a suitable user interface.

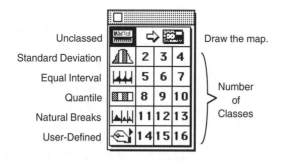

Figure 3.3   Data classification menu palette from a computer mapping program.   Different classifications of a data set can be selected by clicking on each box in the menu palette. The options along the left side control the method of data classification. The numbers indicate the number of classes and the box with the arrow draws the map on the screen.

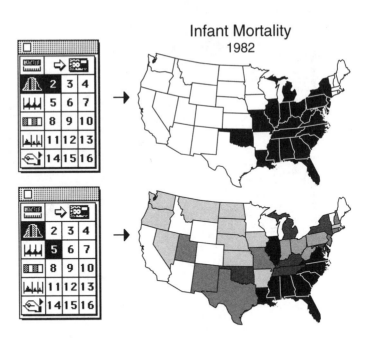

Figure 3.4   The effect of a different number of data categories is depicted on maps of infant mortality in the U.S. The classification for both maps is based on the standard deviation.

3.4 THE ANIMATED MAP

Animation is a "graphic art that occurs in time" (Baecker and Small 1990). It is a dynamic visual statement that evolves through movement or change in the display. The most important aspect of animation is that it depicts something that would not be evident if the frames were viewed individually. In a sense, what happens between each frame is more important than what exists on each frame.

The advantages and feasibility of cartographic animation were described as early as 1959 by Thrower (Thrower 1959, 1961), who viewed its potential from the perspective of film. The computer was soon being used to create the individual frames (Cornwell and Robinson 1966). In the years since, there have been few examples of cartographic animations, due largely to the complexity of their creation, presentation, and cartographer's fixation on the printed map (Campbell and Egbert 1990; Karl 1992). Exceptions include cartographic animations to depict the growth of a city (Tobler 1970), traffic accidents (Moellering 1972, 1973a, 1973b), population growth in an urban region (Rase 1974), and three-dimensional cartographic objects (Moellering 1980a, 1980b).

Recent contributions in cartographic animation have attempted to formulate a conceptual foundation for the use of animation in cartography. Monmonier (1990) has proposed a scripting mechanism to direct the display of a map series, and DiBiase et al. (1992) outline a series of dynamic variables for cartographic animation. However, a lack of software tools that effectively automate the cartographic animation process has probably been the major obstacle to the widespread implementation of animation in cartography.

The animation of maps has been predominantly associated with the representation of change over time. Examples of temporal animations would include changes in per capita income, the increase in population density, or the diffusion of a farming method such as irrigation. Cartographic animations are useful for other purposes as well, such as depicting the deformation caused by a map projection (Gersmehl 1990), a three-dimensional surface (Moellering 1980a, 1980b), or the classification of data (Peterson 1993). Such *non-temporal* uses of animation in cartography may evolve into the major application of the technique.

3.4.1 Temporal Animation

In practical terms animation is creating the illusion of change by rapidly displaying a series of single frames, as with film or video (Roncarelli 1988). A common example would be the movement of a cartoon character. Movement can also be interpreted as the change in the perspective of the observer as the figure remains still.

In cartography, animation is usually defined as the depiction of change through time. The maps in Figure 3.5 are selected frames of an animation that depicts the change in the percent of population over 65 in the United States. Although the data are recorded at 10 year intervals by the U.S. Census Bureau, data values can be interpolated on a yearly or subyearly basis to create more frames (see Chapter 11). The objective of this type of animation is to show change over time.

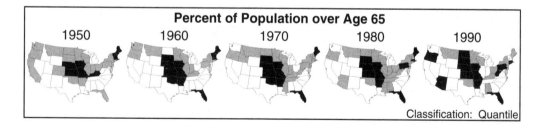

Figure 3.5   Individual frames of a temporal cartographic animation. Changes are more gradual if the data is depicted on a yearly or sub-yearly basis. All maps show three classes with the darker shadings indicating a higher percentage of population over 65.

3.4.2 Non-Temporal Animation

Other types of cartographic animation consist neither of movement nor change through time. The maps in Figure 3.6 are individual frames from a generalization type of cartographic animation. The maps depicted in the animation change from two to six classes and show the effect of the number of data categories on a mapped distribution.

The effect of data classification can be observed with a classification animation. Here each frame of the animation depicts a different classification scheme. A number of different statistical and nonstatistical methods exist for classifying quantitative data (e.g., standard deviation, natural breaks; see Dent 1993). Figure 3.7 depicts the individual frames of a classification animation. The viewing of a classification animation can present the variety of classification options quickly and provide a less misleading view of the data than simply relying on one map.

Another type of animation would be the depiction of a spatial trend. A spatial trend is evident when examining a series of related variables. For example, the percentage of population in age groups (5–13, 19–24, 45–54 years of age) will usually show a clear regionalization in a city, with the older populations closer to the center and younger populations nearer the periphery. An animation of age groups from younger to

Figure 3.6   Selected frames of a generalization animation. The map at the far left with two classes is the most generalized depiction of the data. The remaining maps depict the data with decreasing levels of generalization — three, four, five, and six classes respectively. The darker shadings indicate a higher percentage of births to mothers under age 20.

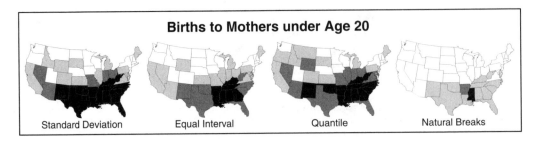

Figure 3.7   Selected frames of a classification animation. All maps depict the data using four classes with the darker shadings denoting states with a higher percentage of births to mothers under 20 years of age.

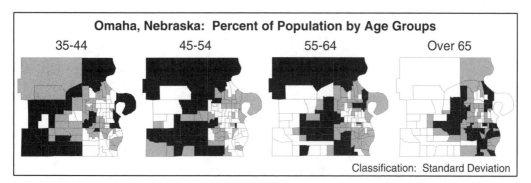

Omaha, Nebraska: Percent of Population by Age Groups

| 35-44 | 45-54 | 55-64 | Over 65 |

Classification: Standard Deviation

Figure 3.8   Selected frames of a spatial trend animation. The objective of this type of animation is to show a spatial trend of a series of related variables. Omaha, Nebraska, developed along the Missouri river, which delimits the city on the east. The older parts of the city correspond to the smaller census tracts in the eastern part of the city. The newer parts of the city (where the younger people live) are further to the west and north.

older depicts high values "moving" from the periphery of a city to the center (Figure 3.8). Variables such as income and housing valuation depict similar geographic trends.

A number of different types of non-temporal animations are possible with maps. Probably the most widely used is the "fly-through." Moellering (1980a, 1980b) showed how an animation could be made of a three-dimensional object by moving around it in time. The technique has been expanded by combining a digital image of the earth and an elevation model. A large number of oblique views are then constructed to simulate flying through a terrain. The method was demonstrated in *LA: The Movie* created by the Jet Propulsion Laboratory. Four frames of the movie are shown in Figure 3.9.

The possibilities of cartographic animation seem endless. Given the number of different applications, cartographic animation may best be defined as the depiction of *change* through the presentation of a series of maps in quick succession. The growth of a city or the flow of a jet stream show change in position. *Temporal animations* depict change through time. *Non-temporal animations* show change that is caused by factors other than time.

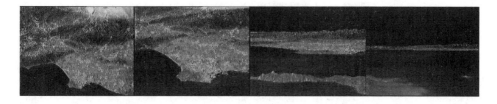

Figure 3.9   Five frames from a production by the Jet Propulsion Laboratory called *LA: The Movie*. The fly–through animation was created by combining satellite imagery with digital terrain elevation data.

3.4.3 Animation Variables

A review of the different animation variables helps to show the potential application of animation in cartography. The variables of animation include both graphical manipulations and sound. Sound can be used to accentuate an animation. For example, a change in pitch could accompany a "cartographic zoom" with the pitch getting higher as one zooms in. The graphical variables of animation include (after Hayward, p. 9):

1. *Size*. The size of an area on a map may be changed to show changes in value. For example, the sizes of countries are made proportionally larger or smaller to depict the amount of oil or coal reserves. An animation can be used to transform the map of oil reserves into the map of coal reserves to show the differences in location of the reserves.

2. *Shape*. An area on a map can be made to change in shape. The shape (and size) of Greenland varies as a result of the influence of a map projection. An animation can be used to blend between the two shapes to accentuate the effect of the different projections.

3. *Position*. A dot is moved across the map to show change in location. For example, the center of population for the United States has moved consistently to the west, and more recently to the south. An animation can be used to depict this movement through time.

4. *Speed*. The speed of movement varies to accentuate the rate of change as, for example, with an animation that depicts the movement in the center of population for the United States.

5. *Viewpoint*. A change in the angle of view, could be used to accentuate a particular part of the map as part of an animation. An animation of population change in the United States may use a viewing angle that focuses attention on the western and southern states where significant increases in population have occurred.

6. *Distance*. A change in the proximity of the viewer to the scene, as in the case of a perspective view. In cartography the distance variable may be interpreted as a change in scale.

7. *Scene*. The use of the visual effects of fade, mix, and wipe to indicate a transition in an animation from one subject to another.

8. *Texture, Pattern, Shading, Color*. Graphic variables that may depict a change in perspective for a three-dimensional object. These may also be used to "flash" a part of the map to accentuate a feature.

3.4.4 Animation as an Exploratory Tool

Animation can be used as an exploratory tool to detect similarities or differences in distribution within a series of maps. This is especially possible when one can interactively access the individual frames in an animation and quickly switch between individual maps or map sequences.

Cartographers have adapted the methods of visualization, many developed within statistics, to the display of maps. One of these methods is called *reexpression* (DiBiase, et al. 1992). The term denotes an alternative graphic representation that results from a transformation of the original data. Three types of reexpression are proposed for cartographic use: brushing, reordering, and pacing:

> **Brushing**: The interactive subsetting of data as in the selection of data values within a scatterplot. For maps it is proposed that points be linked to their geographic location on the map (Monmonier 1989). One could also "brush" across the map to select the corresponding points in the scatterplot.

> **Reordering**: The order of scenes in a time-series animation is usually from beginning to end. Reordering involves the presentation of the scenes in a different order, usually according to an attribute. DiBiase, et al. (1992) give the example of depicting earthquake events. A typical time-series animation would depict earthquake events through time. Another approach would be to order the frames by the number of deaths caused by the earthquake. In this way an emphasis is placed on a measure of earthquake severity.

> **Pacing**: Pacing refers to varying the duration of scenes. Once again, using the earthquake example, DiBiase, et al. (1992) propose that the duration of the scene be proportional to the magnitude of the earthquake or the number of deaths.

3.5 ANIMATION AND THE USER INTERFACE

One of the major trends in computer-user interface design is the incorporation of animation. Animated displays are already being used as part of the user interface. For example, the opening of a folder on the Macintosh computer is accompanied by the display of a box that emerges from the outline of the folder and then becomes progressively larger until it opens as a window. Another example is the use of animated icons that indicate the function that will be performed if the icon is selected. Animated icons are used to distinguish between the available options in programs like Adobe Premiere.

In addition to indicating "what is happening," as with the opening of a folder, or "what can be done," as with the use of animated icons, Baecker and Small (1990, 257) suggest that animation can be used to "review what has been done," to "show what cannot be done," to "guide a user as to what to do" or "what not to do." They suggest seven uses of animation for the user interface:

USE OF ANIMATION	DESCRIPTION
Identification	*What is this?*
Transition	*From where have I come to where am I going?*
Choice	*What can I do now?*
Demonstration	*What can I do with this?*
Feedback	*What is happening?*
History	*What have I done?*
Guidance	*What should I do now?*

(After Baecker and Small 1990.)

Animations can serve an important role in the computer-user interface. It can help us to review the past, understand the present, and describe the future. The use of animation in cartography has the same potential.

3.6 SUMMARY

The incorporation of interaction and animation into cartography is an attempt to better utilize the mental imaging capabilities of the human mind and the dynamic nature of human information processing. The computer is an important tool in adding both elements to the display of maps. The ideas presented in this chapter represent only a fraction of the possibilities presented by a more dynamic cartography.

The types of interaction with a map can be categorized as either graphic or attribute-based. Interacting with the graphic could involve roaming through a map display or making the map larger. Attribute-based interactions involve the manipulation of the features through a separate database. An example here would be the display of a particular street by clicking on the street name. A cartographic zoom, in which more features are added as the map is made larger, would also be an example of an attribute-type interaction.

The animated map can be defined as a cartographic statement that occurs in time. It is based on the human sensitivity for the detection of movement. The objective of cartographic animation is to make something apparent that would not be visible if the maps were viewed individually. A basic distinction can be made between temporal and non-temporal cartographic animation. Temporal animations are limited to the display of change over time. Non-temporal cartographic animations include all other forms of animation in cartography.

3.7 EXERCISES

1. What is the importance of interaction in learning? Why has interaction been neglected in cartography? What effect has the lack of interaction had on map use?

2. A map on paper depicts a part of the earth at an instant in time. Can a single frame of a cartographic animation be defined in the same way?

3. Is the definition of the word "map" a function of the current technology? We now refer to a single depiction on paper as a map. Will we eventually use the same word to refer to an entire series of maps in an animation?

3.8 REFERENCES

CAMPBELL, C. S. AND EGBERT, E. L. (1990) "Animated Cartography: 30 Years of Scratching the Surface." *Cartographica* 27, no. 2: 24–46.

CORNWELL, B. AND ROBINSON, A. (1966) "Possibilities for Computer Animated Films in Cartography." *Cartographic Journal* 3, no. 2: 79–82.

DENT, B. (1993) *Cartography: Thematic Map Design.* Dubuque, IA: Wm. C. Brown.

DIBIASE, D., MACEACHREN, A., KRYGIER, J., AND REEVES, C. (1992) "Animation and the Role of Map Design in Scientific Visualization." *Cartography and Geographic Information Systems* 19, no. 4: 201–214, 265–266.

GERSMEHL, P. J. (1990) "Choosing Tools: Nine Metaphors of Four-Dimensional Cartography." *Cartographic Perspectives* no. 5: 3–17.

HAYWARD, S. (1984) *Computers for Animation.* London: Focal Press.

KARL, D. (1992) "Cartographic Animation: Potential and Research Issues." *Cartographic Perspectives* no.13: 3-9.

MACCHORO II WITH MAP ANIMATION. (1989) *Image Mapping Systems*, Omaha, NE.

MCMASTER, R. B. AND BUTTENFIELD, B. P. (eds.) (1991) *Map Generalization: Making Rules for Knowledge Representation.* New York: John Wiley.

MOELLERING, H. (1980a) "The Real-Time Animation of Three-Dimensional Maps" *The American Cartographer* 7: 67–75.

MOELLERING, H. (1980b) "Strategies for Real Time Cartography." *Cartographic Journal* 17: 12–15.

MOELLERING, H. (1973a) "The Computer Animated Film: A Dynamic Cartography." *Proceedings, Association for Computing Machinery*, 64–69.

MOELLERING, H. (1973b) "The Potential Uses of Computer Animated Film in the Analysis of Geographical Patterns of Traffic Crashes." *Accident Analysis and Prevention* 8: 215–227.

MOELLERING, H. (1972) "Traffic Crashes in Washtenaw County Michigan, 1968–70." *Highway Safety Research Institute*, University of Michigan.

MONMONIER, M. S. (1990) "Strategies for the Visualization of Geographic Time-Series Data." *Cartographica* 27, no. 1: 30–45.

MONMONIER, M. S. (1989) "Geographic Brushing: Enhancing Exploratory Analysis of the Scatterplot Matrix" *Geographical Analysis* 21, no. 1: 81–84.

PETERSON, M. P. (1993) "Interactive Cartographic Animation." *Cartography and Geographic Information Systems* 20, no.1: 40–44.

RASE, W. D. (1974) "Kartographische Darstellung Dynamischer Vorgänge in Computergenerierten Filmen." *Kartographische Nachrichten* 6: 210–215.

RONCARELLI, R. (1988) *A Computer Animation Dictionary*, New York and Berlin: Springer-Verlag.

THROWER, N. (1961) "Animated Cartography in the United States." *International Yearbook of Cartography* 1: 20–29.

THROWER, N. (1959) "Animated Cartography." *Professional Geographer* 11, no. 6: 9–12.

TOBLER, W. (1970) "A Computer Movie Simulating Urban Growth in the Detroit Region." *Economic Geography* 46: 234–240.

Part II

TOOLS OF INTERACTIVE AND ANIMATED CARTOGRAPHY

The computer offers a variety of tools for the creation of interactive and animated maps. The tools may be existing programs in computer mapping, graphic design, image processing, multimedia, and computer animation. The tools are also computer languages for scripting or programming. The following six chapters examine these tools.

4

Computer Mapping

4.1 INTRODUCTION

The development of computer mapping is closely related to that of computer graphics. The first recorded use of a computer to create a graphic depiction was in 1952 on a Whirlwind computer at MIT. However, early computers were generally used for other purposes, and the specialized equipment needed for computer graphics was expensive. As a result, computer mapping did not emerge until the early 1960s and it was the 1970s before the quality of the graphics produced by the computer became acceptable for cartographic applications.

The computer has had a profound effect upon cartography, leading first to a reevaluation of the artistic or subjective aspects of map creation and ultimately to a questioning of the medium of cartography. This chapter first examines how maps are created with the computer and the historical development of computer mapping. The current state of computer mapping is then presented by reviewing contemporary programs for the microcomputer.

4.2 SPATIAL DATA

The computer works with numbers. To create a map with a computer requires that the map be expressed digitally. Two basic approaches have emerged for encoding the map — raster and vector. The approaches differ in how space is encoded. In the encoding of an area, for example, the vector approach defines the area with bounding lines. In the raster approach, the area or its outline is encoded with individual cells or pixels (term derived from "picture element").

4.2.1 Spatial Data Encoding

Figure 4.1 depicts the encoding of a polygon with the vector approach. The vertices of the polygon are converted to x, y coordinates with two-dimensional Cartesian coordinates. The Cartesian coordinate system, devised by the French mathematician Rene Descartes (1596–1650), is based on perpendicular x and y axes. A vector is a line that points in a certain direction and can be defined with a pair of x, y coordinates.

In raster encoding, grid squares approximate the outline of the polygon or its interior (Figure 4.2). The individual grid squares or pixels translate into an array of numbers (stored as a one-dimensional list of numbers by the computer). Smaller grid squares improve accuracy, but the volume of data collected increases geometrically.

Both the vector and raster methods of map encoding have advantages for certain applications. The vector approach is best for the encoding of maps that are composed primarily of lines. The raster method, having its origins in the encoding of pictures, is best used to represent a surface that varies continuously in space such as elevation or temperature. It also has advantages in analytical operations such as neighborhood analysis, and is therefore a common method of map encoding in more analytically oriented geographic information systems.

The relative advantages of each approach are highlighted in the United States Geological Survey's (USGS) methods of encoding the digital versions of its topographic maps. A single map is represented by two separate digital products. The Digital Line Graph (DLG) is a vector-formatted file that includes the transportation network, hydrography, and political boundaries. The Digital Elevation Model (DEM) represents

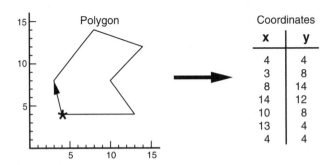

Figure 4.1   An example of vector encoding. The vertices of a polygon are converted to a series of x, and y coordinates for representation by the computer.

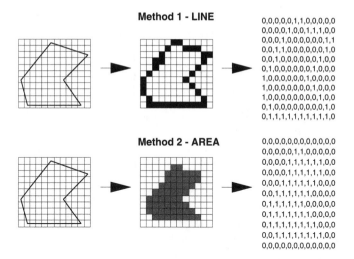

Figure 4.2   Two examples of raster encoding. In the first method a polygon is represented in a grid by encoding its outline. In the second approach the interior grid cells are encoded.

elevation in a raster format. A decisive factor in representing the elevation data with the raster approach is the enormous digital files required for digitizing the contour lines. In addition, the manipulation of the elevations within a grid is easier than the alternative encoding of the contour lines with x, y coordinates.

4.2.2 Spatial Data Structure

The conversion of a map into x, y coordinates or a grid can result in the creation of extremely large files. One of the major concerns in both cartography and geographic information systems is the efficient encoding and storage of maps. One source of inefficiency is the encoding of complete polygons with individual vectors (Figure 4.3). The method results in a duplication of points between adjacent polygons.

The arc-node system (Douglas and Peucker 1973) leads to a more efficient storage of coordinates. In this approach the polygon is subdivided into line segments called *arcs*. The individual arcs are delimited by points called *nodes*. A table is used to describe which arcs constitute a particular polygon (Figure 4.4). When a map is produced, polygons are subsequently reassembled from the table. This method reduces the storage of x, y coordinates by almost one half by eliminating the duplication of points between adjacent polygons. The arc-node data structure also facilitates the overlay of two maps and such other operations as redistricting (the redrawing of political boundaries to assure equal representation). These operations are made possible by defining the relationships between arcs and the polygons they define. The definition of these relationships as well as those between features is referred to as topology.

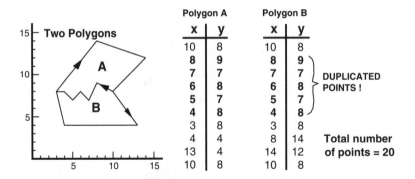

Polygon A		Polygon B		
x	**y**	**x**	**y**	
10	8	10	8	
8	9	8	9	
7	7	7	7	DUPLICATED
6	8	6	8	POINTS !
5	7	5	7	
4	8	4	8	
3	8	3	8	
4	4	8	14	**Total number**
13	4	14	12	**of points = 20**
10	8	10	8	

Figure 4.3   Encoding adjacent polygons stores duplicate points. In this example the two polygons — A and B — share five coordinates that are included in the definition of each polygon.

Software that is used to digitize maps can be set to avoid digitizing the same arc twice. If duplicate arcs are defined by having the same nodes, the three arcs that represent the two polygons in Figure 4.4 are identical. To distinguish between the three arcs, a pseudo-node (false node) would have to be inserted in two of the three arcs, resulting in a total of five arcs for the two polygons. This may increase the total amount of storage but is still less than that used for polygon encoding.

The raster approach also has a number of coding schemes for more efficient data storage. One method, referred to as run-length encoding (RLE), conserves space by encoding "runs" of numbers by their length (Figure 4.5). A "run" is defined here as a set of identical values along a row. The method can significantly reduce the size of the resultant file if there is much repetition in the raster image. If, however, there is no repetition and every pixel has a value different from its predecessor, the size of the run-length encoded file will be twice that of the original.

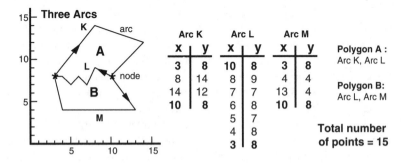

Arc K		Arc L		Arc M		
x	**y**	**x**	**y**	**x**	**y**	**Polygon A :**
3	8	10	8	3	8	Arc K, Arc L
8	14	8	9	4	4	
14	12	7	7	13	4	**Polygon B:**
10	8	6	8	10	8	Arc L, Arc M
		5	7			
		4	8			**Total number**
		3	8			**of points = 15**

Figure 4.4   Arc-Node encoding of polygons. In this example the two polygons are represented with three arcs — K, L, and M. Polygon A is composed of arcs K and L, while polygon B consists of arcs L and M. The total number of coordinates needed to represent the two polygons is 15.

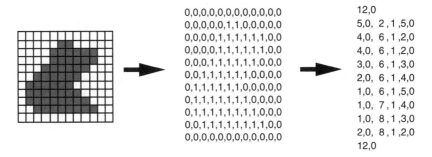

Figure 4.5 Run-length encoding. The method encodes runs of identical values along the horizontal direction. The technique can result in considerable savings in storage.

4.2.3 Spatial Data Manipulation

Once the map is encoded in a digital form, a number of manipulation techniques are possible. For example, with the vector approach, creating a map at a larger size would simply involve multiplying all of the x, y coordinates by the corresponding factor. Figure 4.6 presents the calculation that would be involved for increasing the size of the map by a factor of 1.6. A critical part of the calculation is the subtraction of the minimum x and y values from each coordinate before multiplication by the scaling factor. The minimum x and y values are then added again so that the scaled figure is placed in the same position relative to the origin. The addition of a constant to the coordinates is called *translation* and results in a movement of the figure relative to the origin.

A number of other manipulations are possible with the coordinates. For example,

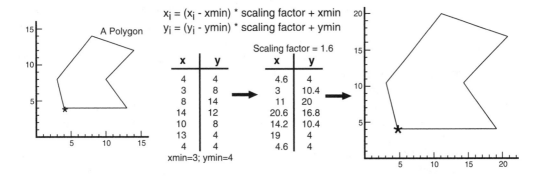

Figure 4.6 Scaling of a polygon. To scale a set of coordinates the difference between each coordinate and the minimum x or y is multiplied by the scaling factor. The minimum x and y values are then added to the scaled value to maintain the same distance from the origin.

the rotation of coordinates is a relatively simple process. The formula for rotation is:

$$x_i = x_i * \cos(\text{radang}) + y_i * \sin(\text{radang})$$
$$y_i = -x_i * \sin(\text{radang}) + y_i * \cos(\text{radang})$$

where, x_i is the set of x coordinates, y_i is the set of y coordinates, and radang is the angle of rotation expressed in radians. The conversion of this formula into a computer program would be as follows:

```
u = cos (radang)
v = sin (radang)
for i = 1, n
        tempx = x(i) * u + y(i) * v
        y(i) = -x(i) * v + y(i) * u
        x(i) = tempx
next i
```

where, u and v store the result of the cos and sin transformation of the angle, n is the total number of coordinates, and tempx stores the result of the rotated x coordinate. It is necessary to store the value of the x coordinate in tempx so that the calculation of the rotated y coordinate can use the original, unrotated value of x.

4.3 THE DEVELOPMENT OF COMPUTER MAPPING

Computers were first built as calculators in the late 1940s to perform engineering and scientific calculations, but experienced their biggest growth in the field of information processing beginning in the early 1950s. Initially the demand for information processing outweighed that of any graphic applications. As a result, computer hardware evolved without graphic capabilities in the 1950s and throughout much of the 1960s.

The development of computer mapping was dependent upon computer graphics peripherals. Yet, as Goodchild (1988, 315) pointed out, the application of computer graphics to computer mapping lagged behind the introduction of computer graphics hardware. This delay was a result of the level of sophistication required for programming. In comparison to mapping, the creation of graphs and charts is a relatively simple programming task. In addition, computer mapping was not a major market for computer graphics hardware, and early peripherals were very expensive. It was sometimes necessary to adapt to available equipment, however inadequate that equipment was for cartographic applications.

The 1970s witnessed a revolution in computer graphics as a number of different graphic peripherals were introduced or became more affordable. In the first part of the decade these peripherals followed a vector approach for display. By the end of the decade the peripherals were based on a raster approach. But the origins of computer mapping go back to the 1960s with a computer peripheral that was not designed for graphic applications at all.

4.3.1 Line-Printer Graphics

Computer mapping was initially performed with line-printers, a standard output device on early mainframe computers. Instead of printing one character at a time (as with the typewriter) the line-printer sets an entire line of text onto paper at once. The printer outputs a page of text in a matter of seconds.

The line-printer was designed for the rapid printing of text, not for computer graphics. It was discovered, however, that one could control the printer to overprint each line to produce shadings. The overprinting of characters such as O, X, H, and I created a relatively dark shading. By modifying the letters used for overprinting, a series of shadings could be created between white and black (Figure 4.7).

A number of computer mapping programs used the line-printer for output. The most widely known was SYMAP, distributed by the Harvard Laboratory for Computer Graphics and Spatial Analysis (Cerny 1972). The program produced different types of maps including choropleth and isarithmic maps, and introduced some important interpolation algorithms (Shepard 1968; for a further discussion of the program, see Monmonier 1982). Maps produced by the line-printer were generally crude in appearance and were criticized by more traditional cartographers (Liebenberg 1976). But programs like SYMAP demonstrated the analytical potential of computer mapping.

Modifying a line-printer by substituting symbols such as " · " and " • " in place of the regular characters, Brassel (1972) also demonstrated the graphic potential of the technology. The modified printer produced a more even gradation of shadings. Brassel used the prototype printer to produce hill-shaded maps (Figure 4.8). This type of map had previously only been created by hand with a graphics tool called an air-brush. Brassel effectively demonstrated that even this highly artistic form of cartography could be computerized.

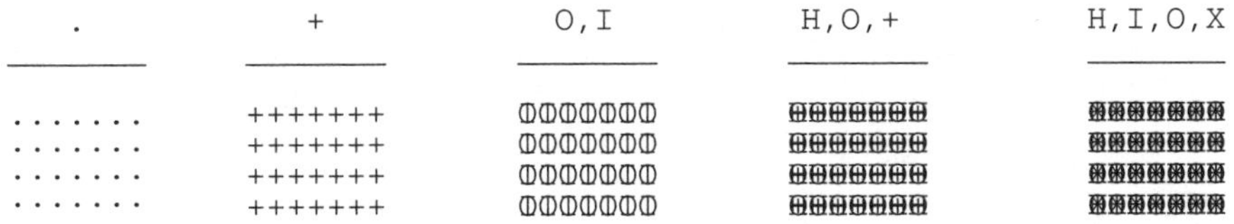

.	+	O,I	H,O,+	H,I,O,X
....	+++++++	⓪⓪⓪⓪⓪⓪⓪	▦▦▦▦▦▦▦	▦▦▦▦▦▦▦
....	+++++++	⓪⓪⓪⓪⓪⓪⓪	▦▦▦▦▦▦▦	▦▦▦▦▦▦▦
....	+++++++	⓪⓪⓪⓪⓪⓪⓪	▦▦▦▦▦▦▦	▦▦▦▦▦▦▦
....	+++++++	⓪⓪⓪⓪⓪⓪⓪	▦▦▦▦▦▦▦	▦▦▦▦▦▦▦

Figure 4.7   Early computer mapping programs used the computer line-printer for output. Shadings were created by overprinting characters. The method was used to create choropleth and shaded isarithmic maps.

Figure 4.8   Example of a hill-shaded map created by a specially modified line-printer. The normal alphabetic characters in the printer were replaced by geometric symbols. (After Brassel 1972.)

4.3.2 Vector Graphics

The introduction of vector graphic peripherals represented an important development in computer mapping because the vector approach more closely approximated manual methods of map drafting, making computer mapping more acceptable to cartographers. A number of different vector graphic peripherals were available by the early 1970s.

Coordinate Digitizer

The coordinate digitizer made the input of maps possible. The tablelike device incorporates a magnetic grid. By manually positioning a cursor over the intended point and pressing a button, the point is recorded as an x, y coordinate pair (Figure 4.9). Digitizing a river, for example, would involve following the line on the map and recording a point at which the line changes direction. A series of points would define a line or a polygon. Digitizing a map in this way is a tedious process but is still the primary method of map encoding. The major advantage of digitizing over other more automated methods of map input is the control that the operator maintains over the digitized product.

Vector Refresh Display

The first terminal to display a vector-defined graphic was the vector refresh or calligraphic display. The vector refresh terminal was a cathode ray tube (CRT) using varying analog voltages to control the deflection of an electron beam directed toward the display (Figure 4.10). After being "excited" by the electron beam, the phosphor on the back surface of the screen emitted light for a very short time, about $\frac{1}{30}$ of a second. The required refresh rate of 30 per second limited the number of vectors that could be displayed. If more vectors were displayed on the screen than could be updated in $\frac{1}{30}$ of a second, the screen flickered as parts of the display faded out. An advantage of the display was that the picture could be dynamically altered because the picture was being constantly redrawn. The minimum cost of the terminal was $15,000 (Machover 1980) and, because of their high price and the "flicker" problem, its utility in computer mapping was limited.

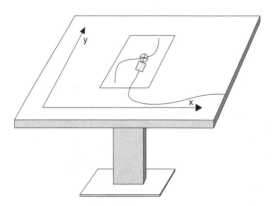

Figure 4.9   Coordinate digitizer. The table-sized device incorporates a magnetic grid. A pointing device or "puck" is used to position a cursor. Clicking on the puck outputs an x, y coordinate pair to a computer, where it is scaled and stored in a file.

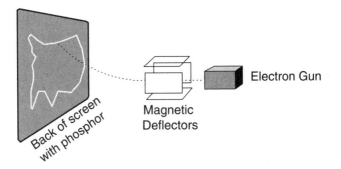

Figure 4.10   Design of a vector refresh terminal. The terminal used magnetic deflectors to control the direction of an electron beam that was directed toward the display. The electron beam excited a phosphor material on the back of the screen that remained illuminated for a fraction of a second. The terminal would continuously redraw the graphic to maintain the image on the screen. If the time it took to redraw the image exceeded the time of phosphor illumination, the image began to flicker as parts of the graphic began to fade.

Storage Tube Display

The storage tube was the solution to both the flicker problem and the cost of the vector refresh display. Similar in design to the vector refresh display, this graphics terminal used a low-energy flood of electrons directly behind the display to keep individual phosphor particles illuminated once they had been activated by the high-energy electron beam. The limitation of the technology was in updating or changing the graphic display. The only way to erase any part of the display was to erase the entire screen and redraw the graphic. The terminal represented a low-cost way of providing a graphic workstation at remote sites (Cook 1980, 42). The cost of models such as the Tektronix 4010 fell to less than $4,000 by the end of the 1970s. Throughout this decade the storage tube was the major form of soft-copy computer mapping.

Digital Incremental Plotter

The third vector output device is the digital incremental plotter. Available in a wide variety of configurations, the plotter essentially moves a marking device such as a pen across a medium such as paper or mylar (Figure 4.11). There are two main types of plotters. Flatbed plotters have a stationary drawing medium with a marker that is moved. Drum plotters have a drawing medium that is moved along one axis, usually the x, while the marker is moved along the other.

The resolution of plotter movements is as little as 0.0003 of an inch (0.000762 cm). However, the time required to produce a map on paper is considerable. Even with pen movement speeds of 40 inches (101.6 cm) per second, several hours might be required to "plot" a complex map. At these pen movement speeds the plotters

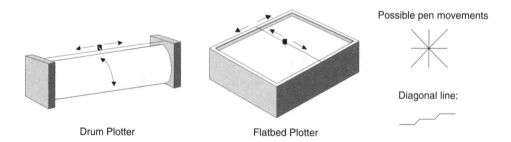

Figure 4.11   Drum and flatbed plotters move a drawing device across a medium such as paper or mylar. Plotter movements are controlled by two motors, one each for the x and y directions. This limits the plotter to drawing a line in one of eight directions.   Diagonal lines that are not at an increment of 45° are drawn with a series of 0° and 45° lines.

would frequently break down. The initial plotters were also expensive. A plotter such as the CalComp 960 cost more than $50,000. The plotter was the major form of hard-copy output throughout the 1970s and early 1980s and is still used today.

4.3.3 Raster Graphics

The major limitation of both computers and graphics peripherals in the 1960s and 1970s was computer memory. The introduction of the dynamic random access memory (DRAM) semiconductor had a major impact on graphic terminal technology. Computer memory could now be installed within the terminal to store and even help manipulate the graphic image. The DRAM semiconductor was especially important to the development of raster-based terminals, and this technology has since dominated computer graphics input and output. The following are examples of raster terminal technology.

Scanner

Scanners were already in existence in the 1960s. The typical design was a rotating drum that held the document and a photosensitive head that moved across the drum. Drum scanners remained fairly expensive (over $100,000 each), and their use was initially limited. Scanners have recently become much less expensive (under $500) and are a common peripheral with microcomputers. Hand-held units are now available and can be used to capture a picture or a graphic in both black and white and color.

The initial digitizing of a map with a scanner is much faster than with a coordinate digitizer. After a map has been scanned, it can be vectorized through a raster-to-vector conversion process. Attributes can then be attached to the individual features in the map. However, the scanning of maps is not a simple process. Automatic vectorization requires a larger workstation computer, and the assigning of attributes is still a time consuming process. Microcomputer programs are available to perform a limited type of vectorization, but the programs are as yet not sufficiently sophisticated to vectorize maps properly (see Chapter 5).

Within institutions that digitize a large number of maps, the scanning of maps has become a more viable alternative to vector digitizing, especially for maps that can be obtained in individual layers (e.g., street, water features and political boundaries available as separate maps). It is very common to redraft a map by hand before it is scanned so that extraneous details can be removed.

Raster Display Terminal

Bit-mapped black and white displays represent the simplest implementation of the raster graphics display technology. For example, the screen of the original Apple Macintosh has a resolution of 512 x 342 pixels. One bit of memory, which can be either on or off, is used to control the state of each pixel on the screen — black or white. A total of 175,104 bits (512 x 342), 21,888 bytes (1 byte = 8 bits), or 21.375 KB (1 KB = 1024 bytes) are needed to "map" the display. This digital "video" memory is converted to an analog signal for purposes of display. This conversion occurs approximately 60 times a second.

Color raster graphic terminals use variations in the intensity of the three primary additive colors — red, green and blue — to display up to 16,777,276 colors, although usually only a smaller number of colors can be displayed at one time. A typical raster graphics terminal or microcomputer uses three bytes for each pixel, one for each of the primary colors (Figure 4.12). One byte can represent 256 digital values (2^8). Thus 256 different intensities of red, green, and blue can be displayed for each pixel. A combination of the different intensities for the three different primaries produces the different colors, up to 2^{24} colors. Of course, the memory requirements are much greater than for black and white. A 512 x 512 display with three bytes per pixel would require 786,432 bytes or 768 KB. A 1024 x 1024 display would need 3,145,728 bytes, 3072 KB or 3 MB (1 MB = 1024 KB) of memory.

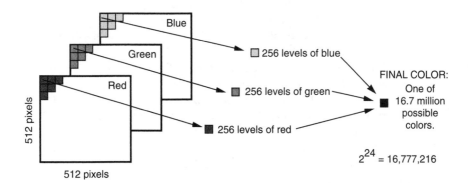

Figure 4.12   Color raster graphic terminal design. Each pixel is represented with three bytes, one byte each for red, green, and blue. Each byte represents up to 256 levels of each color. The possible combinations of 256 levels of each of the three primary colors results in the representation of over 16.7 million possible colors.

Raster Printers

The dot-matrix, ink-jet, electrostatic, laserprinter, and imagesetter are all examples of printers using the raster method for output. A common characteristic of all of these printers is that the graphic is produced line-by-line as the paper is moved across a printing surface that is oriented perpendicular to the direction of paper movement.

The electrostatic plotter was an early alternative to the digital incremental plotter. Specially treated paper moves over a bar that contains tiny wires or nibs that place a charge on the paper. Subsequently a toner is applied to the paper that adheres to the charged areas. A suction device then removes the excess toner. The initial resolution of electrostatic plotters was 100 dots per inch. Color electrostatic plotters are now available with a resolution up to 400 dpi.   These plotters use four passes to apply the three subtractive primary colors — yellow, magenta and cyan — and black separately with the paper being recharged between each pass.

The laser printer works on a similar principle, except that a laser directed by a rotating mirror, is used to charge the paper. The resolution of raster printers is controlled by the medium used for the output. The highest resolution that is possible with paper is about 800 dpi. Imagesetters that print on film have a resolution of up to 3,386 dpi (0.0002953 inch resolution). In general, raster printers are faster and the equipment is more reliable than digital incremental plotters, and have for the most part replaced this vector output device.

4.3.4 The Microcomputer

Of all the hardware developments in computer graphics the most significant has been the microcomputer. As recently as 1975 the microcomputer had not yet been developed. Today the personal computer is an established consumer product sold like any other piece of electronic equipment. Microcomputers are used in the office, in the home, in the laboratory, in the school, on airplanes, and at the beach. Machines that had filled large rooms and required the attention of numerous individuals were reduced to desktop units operated by a single person.

The microcomputer had its beginning with the development of smaller computers in the early 1960s. In 1965 Digital Equipment Corporation (DEC) introduced the PDP-8, the first computer to use integrated circuits. DEC used the term minicomputer to describe the product, and a new market formed. Minicomputers by DEC and other companies such as Data General and Prime were used in scientific laboratories, small businesses, and factories. As the price of these machines fell and their capabilities improved, more and more businesses automated with minicomputers (Shurkin 1984, 279).

It is generally accepted that the first microcomputer was the Altair from MITS Corporation of Albuquerque, New Mexico. The company had started by selling kits for calculators at $99. By early 1974 the price of calculators dropped drastically as companies that produced the processing chips realized that there was a large market for calculators, and started their own production of these devices (e.g., Texas Instruments). MITS responded by designing a computer kit. The kit was released in January 1975 at a price of $397 (Freiberger and Swain 1984, 35). The computer had 16 expansion slots, a CPU board with an Intel 8080 8-bit processor, a memory board, and an input/output (I/O) unit. It did not have a permanent storage unit, so that once the machine was shut-off, whatever information had been entered was lost. The I/O system consisted of a series of switches and tiny flashing lights. The computer only understood machine language — a set of commands understood by the processor and limited to moving data between internal storage registers, storing data in memory, and performing simple arithmetic.

The following two years saw tremendous growth in the microcomputer industry. By the end of 1977 a total of 29 separate companies were involved in the microcomputer business, almost all start-up companies. Most of these microcomputers used the same design, Intel 8080 processor, and operating system (CP/M) as the Altair. Apple Computer, founded by Steven Jobs and Stephen Wozniak, took a different route. In July 1976 the first Apple I was completed, based on the Motorola 6502 processor. The Apple I provided only the essentials, lacking a keyboard, power supply, and case (Freiberger and Swain 1984, 212).

The Apple II, released in mid-1977 and based on the same Motorola processor, was a complete computer, with a keyboard, power supply, the BASIC programming language, color graphics, and a case. Containing only 4K of RAM, the computer cost $1,298 and connected to a television for screen input and output. An important peripheral added by Apple in 1978 was a floppy disk drive that replaced a cassette tape drive for the storage of programs and files (Freiberger and Swain 1984, 213). The microcomputer revolution was under way, fueled largely by important software packages that included text-processing and spreadsheet programs (e.g., WordStar and VisiCalc).

The next major development in the microcomputer industry was the introduction of the IBM personal computer or PC in August 1981. It used the Intel 8086 16-bit processor, relatively standard hardware, and an operating system called MS-DOS that was written by Microsoft, a Bellevue, Washington, company. The PC and all of its clones, based on a series of Intel processors called the 80086, 80286, 80386, 80486 and Pentium, have dominated microcomputer sales ever since.

The last development in the first decade of the microcomputer was the introduction of the Apple Macintosh in 1984. The origins of the Macintosh extend back to 1979 when Steven Jobs visited Xerox's Palo Alto Research Center (PARC), a laboratory funded by Xerox for high-technology research. During his tour at PARC, Jobs saw a demonstration of a computer using a mouse-controlled user interface. The computer had a much better graphics resolution than the Apple II and could display graphical icons. The mouse could be used to select options by pointing to and selecting individual icons. It was an input device conceptually distinct from anything then in use with computers. Jobs set out to create a widely affordable computer with all the technological innovations of the computer he had seen at PARC.

The Apple Macintosh changed the human interface with computers. The computer implemented a consistent and intuitive user interface based on windows, menus and dialogs. The operation of programs could be learned much more quickly, often without reading a manual. The computer popularized the point-and-click user interface, and it spurred the development of similar operating systems for other computers (e.g., Microsoft's Windows for DOS and IBM's OS/2). The Macintosh has also become associated with certain applications, particularly computer graphics and multimedia. Most newspapers, for example, create their maps and graphical illustrations with Macintosh computers.

The development of the PowerPC series of processors, a joint effort by Motorola, Apple and IBM, is helping to merge the Macintosh and PC computers made by IBM. Both Apple and IBM are using the PowerPC processors in their computers. These processors, based on a RISC (Reduced Instruction Set Computing) architecture, have very fast computing speeds. Although both companies still use different operating systems (Apple's Macintosh OS vs. IBM's DOS and OS/2), it is expected that the use of the same processors will eventually lead to a greater compatibility between these computers.

4.3.5 Computer Operating Systems

As important as the development of microcomputer hardware was the introduction of new operating system software. An operating system is the software that acts as the interface between the computer hardware and the user. It is a program that accepts commands and makes the operation of the computer possible. There are different operating systems for computers, three examples of which are: Macintosh OS, MS-DOS/Windows, and UNIX.

The Macintosh operating system (OS) consists of a system file and a program called the Finder. The Finder is the base program that displays the icons of the programs on a disk. Programs, also called an applications, are launched by double-clicking on the associated icon. The Finder also makes it possible to launch a program by double-clicking on a file that was created by a program.

MS-Windows is not an operating system but rather a Finder-like program that runs in conjunction with MS-DOS (Microsoft's Disk Operating System) on IBM-PC and PC-clone computers. It is accessed in MS-DOS through the win command. Windows has implemented many features of Apple's Finder, including icons and pull-down menus. A great deal of MS-DOS and Macintosh software is being rewritten to run under Windows. Two new versions of Windows, Windows NT and Chicago, do not need to run in conjunction with MS-DOS.

UNIX was developed in the late 1960s at AT&T. It was designed as an operating system that would work with different computers. It has become the major operating system for workstation computers. Although it has a cryptic command-line interface, graphical user interfaces such as X-windows, SunView, and Motif have also been developed for UNIX.

4.3.6 Palm-Top Computers

An obvious trend in computer technology has been an increase in processing power and a reduction in hardware size. The latest example of this trend is a hand-held computer that incorporates a stylus interface, commonly referred to as the palm-top. Because of their portability, such computers will influence how maps of the future are used.

The major emphasis in the development of palm-top computers has not been the miniaturization of the hardware but the development of operating system software that make these computers useful. There are currently three major operating systems for palm-top computers: Go Corporation's PenPoint, Microsoft's PenWindows, and Apple's Newton OS. The first two companies are not involved in the production of hardware.

GO Corporation has designed a prototype computer to demonstrate its associated PenPoint operating system. About the size of a photo album, the GO computer is all monitor — there is no keyboard. The liquid crystal display (LCD) screen has a resolution of 100 dots per inch. The PenPoint operating system uses a stylus interface and is specially designed to interpret handwriting. The operating system incorporates the recognition of meaningful gestures such as caret for insert, cross-out for delete, square-bracket for block-marking, and question mark for help.

PenWindows from Microsoft is the most widely used of the stylus interfaces because it runs in conjunction with MS-DOS (a popular operating system for microcomputers). Software has been developed for PenWindows that allows people working with a GIS data base to perform updates in the field by writing onto a map displayed on the screen. A number of computer manufacturers have made computers that work with PenWindows.

Apple's Newton Personal Data Assistant is also based on a stylus interface and includes software for handwriting recognition. The display consists of a combination writing surface and display screen. Included with the computer is software for taking notes and sketching figures. The software converts rough, hand-drawn figures into clean symmetrical shapes that can be subsequently edited. Some models have wireless communications capabilities using infrared light waves.

The development of palm-top computers is still in its infancy. For the presentation of maps, a computer with a larger and higher-resolution screen display may be required (see discussion in Chapter 1). As opposed to the initial development of computers, graphics applications seem to be embedded in the design of palm-tops. The graphics requirements of the palm-top user interface will certainly contribute to the development of graphics applications with these computers.

4.3.7 Computer Mapping Software

A computer can do very little without software. Programs are the codes by which we tell the computer what to do. Early programs for computer mapping were written for specific graphic devices. Terminals from Tektronix and plotters from CalComp were supplied with subroutine libraries that made their control possible from within computer programs. The PLOT-10 library from Tektronix, a maker of storage tube terminals, and the CalComp library from the major producer of plotters were two popular libraries in the 1970s. The PLOT-10 library was more sophisticated, supporting a type of window called a viewport. The initial CALCOMP library did little more than connect points with lines through a PLOT subroutine call. Figure 4.13 presents a small portion of a program that would draw the polygon from the first illustration in this chapter (Figure 4.1). The first "call" statement moves the drawing device to the first point. The third argument in the CALCOMP "call plot" statement controls whether the pen is up or down when moving to the point (3 is pen up; 2 is pen down). A loop is then used to connect the remaining points in the polygon.

```
FUNCTION                  CALCOMP                   TEKTRONIX PLOT-10
_____                  _____                   _____

move to first point       call plot(x(1),y(1),3)    call movea(x(1),y(1))
start loop                do (i=2,7)                 do (i=2,7)
draw line to point          call plot(x(i),y(i),2)    call drawa(x(i),y(i))
end loop                  enddo                     enddo
```

Figure 4.13   Use of a subroutine library to draw a polygon as depicted in Figure 4.1. The first statement calls a subroutine to *move* to the first point. The second statement begins a loop with a variable "i" that begins with 2 and ends with 7. The third statement *draws* a line that connects the 6 x, y coordinates. The last statement ends the loop.

Incompatibilities between graphics libraries led to the development of *device independent* subroutine libraries. The purpose of this type of software was to make programs for computer graphics independent of the graphic output device. This was accomplished through the creation of a graphics meta file, a file that contained all instructions for drawing the graphic. The meta file was subsequently interpreted by a device-specific program to output to a specific device. Three major graphics libraries have been developed: (1) SIGGRAPH Core, developed by the Association of Computing Machinery Special Interest Group in Computer Graphics, (2) GKS, the Graphical Kernel System developed by the European Standards Organization (GKS was originally adopted by the International Standards Organization) and, (3) Phigs, the Programmer's Hierarchical Interactive Graphics System. Although there is as yet no accepted standard for computer graphics, many computer graphics applications on UNIX-based computers are programmed with the PHIGS library. The International Standards Organization (ISO) has now adopted the PHIGS library of graphics routines.

Device-independent libraries make it possible for software developers to support a wide variety of different graphics peripherals. But by concentrating on the common denominator between different devices such libraries do not take full advantage of the graphics capabilities of particular output devices. They also tend to implement their own user interface rather than the interface that is associated with the computer's operating system.

Of greater importance to computer graphics are a variety of de facto file format *standards* that have emerged during the 1980s. The PostScript™ page description language, for example, is recognized as a standard for the definition of graphics and text for output to printers. A graphics format called PICT is used on the Apple Macintosh. A format devised by AutoCAD, a maker of computer-aided design software, is commonly used for the exchange of vector files. The lack of a single standard in computer graphics tends to be wasteful of time and energy. However, standards can inhibit progress and the search for better techniques.

To understand the development of computer mapping software, the three types of computer mapping programs that have emerged since about 1980 need to be examined. The first category consists of statistical analysis programs that have incorporated thematic mapping functions in addition to general statistical procedures. The second type are computer mapping programs specifically written for microcomputers and referred to as "desktop mapping." This category includes programs for (1) data display and analysis, (2) contour and three-dimensional mapping, (3) map projections, and (4) map digitizing. The third category is based on programs called geographic information systems that have a significant data base component.

4.4 STATISTICAL ANALYSIS PROGRAMS

Statistical analysis programs, dating to the late 1960s, were developed on mainframe computers, and retain somewhat of a "card-like" input structure. They are characterized by a large number of statistical functions, including correlation and regression analysis. By the early 1980s graphing and mapping functions were added to the mainframe and minicomputer versions. Microcomputer versions, released by the mid-1980s, often implemented only a subset of the original functions.

SAS/Graph (SAS Institute Inc.)

SAS/Graph is an extension of the Statistical Analysis System (SAS) and consists of a series of graphing and mapping procedures. The program is device-independent, supporting a wide variety of different output devices from plotters to raster graphic terminals. JMP is a more interactive version of SAS designed for microcomputers. SAS/GIS is another extension to SAS that combines geographical data management, analysis, and presentation.

Producing a map with SAS requires two data sets: a map data set and an attribute data set. The map data set contains a set of coordinates that represent the boundaries of the map areas. The attribute data set is the data that are to be mapped. There are three basic procedures to create maps in SAS/GRAPH: GMAP, GCONTOUR, and G3D. The GMAP procedure creates four different types of thematic maps — choropleth, three-dimensional surface, three-dimensional raised block, and graduated bar based on map files in a polygon format. GCONTOUR and G3D produce contour and three-dimensional netted-surface maps based on data in a raster format. An accessory procedure called G3DGRID assists in calculating a raster data set.

A large number of different map files are supplied with the program. The maps are stored in a polygon format. A record in a map file consists of the x, y coordinates, an identification number (country, state, or county code), and a segment number. The segment number identifies individual polygons within a map area. The procedure GPROJECT is for projecting map files (e.g., Alber's Equal Area, Lambert's Conformal, or Gnomonic). The GREDUCE program simplifies map outlines by removing points based on a line simplification algorithm. The procedure GDEVICE provides access to the separate device drivers to direct the output format.

SPSS Graphics and SPSS/PC+ Map *(SPSS Inc.)*

The other major statistical package for mainframe and minicomputers is the Statistical Package for the Social Sciences (SPSS). The mapping procedures in SPSS Graphics were added to the package to create only choropleth and three-dimensional prism maps. The program incorporates map projection algorithms and includes a number of editing capabilities such as the removal of overlapping text labels

SPSS/PC+ Map is a PC version of SPSS that works within the Windows graphical environment. This version supports mapping functions through the MapInfo desktop mapping program, a program described later. A module from MapInfo integrates the SPSS and MapInfo software products to create thematic maps.

SYSTAT *(SPSS Inc.)*

SYSTAT has many of the same statistical functions as SAS and SPSS/PC. The major distinction of this statistical program is that it was originally designed for microcomputers. The program produces choropleth, contour, raised-line, Chernoff face thematic maps, and map projections, including the stereographic, gnomonic, Mercator, and perspective view. The program is available for both the PC and the Macintosh.

4.5 DESKTOP MAPPING

The term *desktop mapping* emerged during the latter part of the 1980s to describe the creation of maps with the microcomputer. The term was a take-off on *desktop publishing,* which was then revolutionizing the publishing industry. The impetus for this major change in document processing was a program called *Aldus Pagemaker* that permitted the layout of multicolumn documents with embedded graphics. By promoting the concept of desktop mapping, hardware manufacturers and software publishers attempted to borrow from the popularity of desktop publishing. However, mapping operations were more complicated than text layout, and the user interface never developed to the same level. Desktop mapping was an important development, and continues to have a great influence on cartography and the development of a more interactive, sometimes even animated, approach to map production and map use.

Desktop mapping has been viewed with considerable apprehension by cartographers. The programs put map-making tools in the hands of people not trained in cartography, resulting in the production of maps that are often incorrect or are of poor graphic quality.   Some programs incorporate safeguards against users making incorrect maps. The maps produced by desktop mapping programs are also designed for use by a smaller audience for whom issues of graphic quality are not as important as they are for maps intended for more general distribution.

A new development in desktop mapping is the integration of mapping functions in spreadsheets and data base programs. Reflective of a general trend toward integrated software, this development is referred to as *embedded mapping* (Schwartz, 1993). Companies, such as Lotus and Oracle, that produce data base programs are incorporating mapping functions. Even producers of word processors are considering the integration of maps. Imbedded mapping represents a further expansion of desktop mapping into existing programs.

Current programs for desktop mapping can be divided into four categories: (1) data analysis and display, (2) contour and three-dimensional mapping, (3) map projections, and (4) digitizing.

4.5.1 Data Display and Analysis

These programs stress the analysis of demographic data. The most common map produced by these programs is the choropleth map. Other possible forms of symbolization include the cartogram, raised-surface, three-dimensional prism, dot, circle, graduated segmented circle and other point-symbol maps. Compared to the statistical analysis programs described above, these programs incorporate only a few statistical functions and have a relatively limited ability to manipulate the data. More complex data manipulation functions will usually need to be performed with a data base application such as DBase or 4th Dimension.

There are a number of programs in this category. Examples include Atlas GIS for Windows (Strategic Mapping), MapViewer (Golden Software), MapInfo (MapInfo Corp.), GeoVista (ALL Systems Design), SUPERMAP (Chadwyck-Healey Inc.), Harvard GeoGraphics (Software Publishing Corp.), GeoQuery (GeoQuery, Inc.) and Scan/US (Scan/US Inc.). The programs are usually supplied with a large amount of map and demographic data. Most incorporate a point-and-click interface.

ArcView® (Environmental Systems Research Institute) is used in the same way as the previous programs but works in conjunction with ARC/INFO®, one of the major programs for geographic information systems. ArcView® integrates the presentation of documents, images, tables, text, graphics, spreadsheets, maps, multimedia, and CAD drawings. Information about features may be retrieved by pointing and clicking on the feature, and searches can be performed on a data base according to specific criteria.

Other programs in this category emphasize specific mapping capabilities. The Map Collection (MapWare) can be used to create a number of different types of maps, including block diagrams, line maps, circles, contour, and choropleth maps. MacChoro II, is limited to choropleth mapping, but incorporates a number of different data classification options and the animation of maps. The data classification options include equal interval, quantile, standard deviation, natural breaks methods and an unclassed and user-defined option. Cartographic animations are created automatically by placing individual choropleth maps in computer memory and playing them back at speeds up to 60 per second.

These programs all make use of demographic data. The major source of this type of data in the United States is the U.S. Census Bureau. This agency undertakes a population census every ten years and regularly releases reports based on sample data collection. Other federal agencies, including the U.S. Department of Health and the U.S. Department of Agriculture, release data by region, by state, or by county.

Companies that market socioeconomic data essentially reprocess data that is distributed by the government, in some cases forming composite statistics. The largest customers for this data are businesses. The types of files that are sold include: crime rates, disease rates, employment characteristics, health-care facilities, housing, retail trade statistics, business activity, and consumer patterns. Companies in this category include CACI, Claritas/NPDC, Equifax National Decision Systems, and Strategic Mapping Inc.

4.5.2 Contour and Three-Dimensional Mapping

Contour and three-dimensional mapping programs use two-dimensional interpolation algorithms to transform scattered x, y, z data points into a grid of data points. Subsequently the isarithmic or contour lines are drawn by threading lines through the matrix of values. Interpolation algorithms calculate values for unknown points based on the surrounding known values. There are a variety of interpolation algorithms, but no single one is appropriate for all types of applications. The most widely used interpolation approach is the inverse distance squared method shown in Figure 4.14.

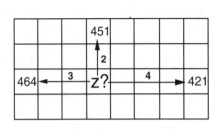

$$z? = \frac{\sum_{k=1}^{m} Z_k / D_k^2}{\sum_{k=1}^{m} 1 / D_k^2}$$

$$\frac{\frac{464}{9} + \frac{451}{4} + \frac{421}{16}}{\frac{1}{9} + \frac{1}{4} + \frac{1}{16}} = \frac{190.61}{0.4236} = 449.96$$

Figure 4.14   The inverse distance squared interpolation formula. Values for unknown cells are computed based on the distance to, and value of, a series of known cells, where, Z_k represents the known grid cell values, D_k is the distance to the unknown cell z?, and m is the number of known data values to use in the calculation — in this case, three.

Other algorithms convert the matrix of elevation values to three-dimensional, netted-surface maps.

Microcomputer programs for contour and three-dimensional mapping include Surfer (Golden Software), Surface III (Kansas Geological Survey), MacGRIDZO (Rockware, Inc.), GeoView (Computer Systemics), LANDview (COMPUneering), and, MacGeos II (SIAL Géoscience). These programs all perform interpolation based on scattered x, y, z coordinates, although the quality of the interpolation varies. The programs also differ in how data can be brought into the program, how many data points can be used, and whether the resultant map can be exported to another program.

Figure 4.15 depicts a contour, a shaded contour, and a netted-surface map created with the Surface III program. The input data consisted of 190 x, y, z points. Values were interpolated for elements of a 26 x 16 grid from the scattered x, y, z points based on the formula in Figure 4.14. Isarithms (contour lines) were threaded through grid to create the contour map. The grid of interpolated values also is the basis of the shaded isarithmic map and the netted-surface map.

4.5.3 Map Projection Programs

A third category of desktop mapping programs produce map projections. These programs do not map data, but simply create alternative views of the earth using a variety of projection formulas. Map projections can be described with formulas that convert the spherical coordinates of latitude and longitude into planar coordinates (x and y) for plotting on paper. For example, the Mercator projection has the formula:

$$x = \mu$$
$$y = \log_e \tan (\pi / 4 + \Omega / 2)$$

where x and y are expressed in two-dimensional Cartesian space, μ is longitude, and Ω is latitude. The Mercator preserves angular relationships (over small areas) but distorts areas, especially in the extreme latitudes. The projection depicts Greenland as being larger than Australia, although the opposite is true (Greenland is farther from the equator than Australia and therefore more distorted in size by the projection formula).

The equal area sinusoidal projection has the formula:

$$x = \mu \cos \Omega$$
$$y = \Omega$$

where μ is longitude and Ω is latitude. This projection depicts the land masses in proper proportion, but exhibits extremes in angular distortion. The programs create both the latitude and longitude grid and the outline of the continents from a data base of coordinates.

There are a number of programs that make maps of the world using different map projections. Relatively inexpensive programs are available for the DOS operating

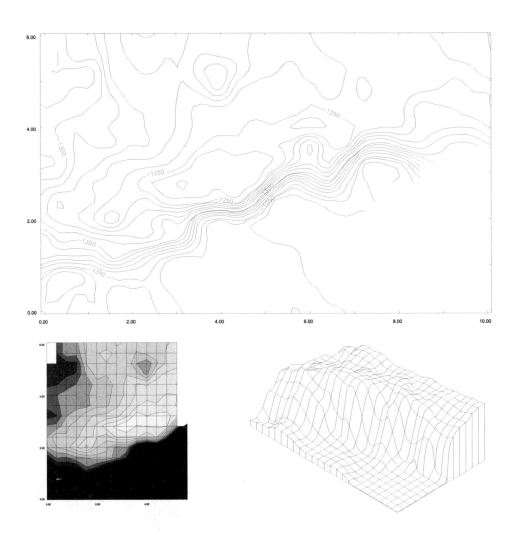

Figure 4.15   Contour, shaded-contour, and netted-surface maps created by the Surface III program. All three maps are based on a grid of elevation values. The program also performs an interpolation of a grid based on a series of x, y, and z coordinates.

system (e.g., World from the University of Minnesota; MicroCam from the AAG Microcomputer Specialty Group). Map projection programs that incorporate a point-and-click interface include Azimuth (Graphsoft) and Geocart (Terra Data, Inc). The latter of these two programs creates over 100 different map projections and nine interrupted types of projections. It also includes drawing tools to add symbols and annotation. All of these programs include a map data base of the world's coastlines, major rivers and lakes, and international boundaries.

4.5.4 Digitizing Programs

Digitizing programs work in conjunction with a digitizer to encode a map as series of x, y coordinates. Digitizing is a tedious process, prone to error, and therefore the programs generally incorporate extensive editing options. The choice of a digitizing program is critical because experience has shown that up to 80 percent of the time required for computer-based analysis of a cartographic data base is consumed in the input and editing phases of a project. GIS programs usually incorporate a digitizing option, but most desktop mapping programs do not.

An example of a program specifically designed for digitizing is Roots (Decision Images). The program supports interactive graphic digitizing and editing with visual feedback, user-defined tolerance values, topology generation, the adding of attribute data, multiple control point registration, graphic output, and conversion to other data formats (Figure 4.16).

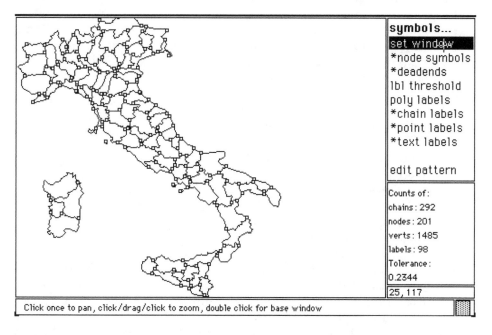

Figure 4.16   Digitizing window from the Roots program. A square symbol marks the nodes for each polygon. The nodes delimit arcs that constitute the individual polygons. (Courtesy of Decision Images.)

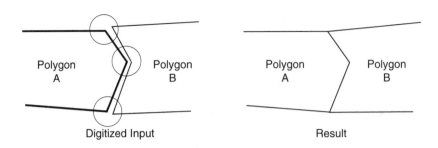

Figure 4.17   The digitizing of polygons. A polygon consists of a series of arcs. Some arcs are shared with other polygons, others are not. The digitizing of arcs is aided by a snap feature that connects an arc to an existing node or arc.

Digitizing programs usually encode polygons by capturing the individual arcs that define the polygon. After digitizing one arc the nodes for subsequent arcs can be made to "snap" to the existing nodes or arcs (Figure 4.17). This snapping occurs within a distance around the node specified by a tolerance value. After digitizing, the polygons are assembled from the series of digitized arcs.

Other programs for digitizing, including DIGITIZE (Rockware, Inc.), do not use an arc approach to digitize features. Features are digitized in the form of points, lines, polygons, and polylines (continuous connected lines). These programs do not create topology and are primarily designed for the input of x, y, z coordinate lists for contour and three-dimensional mapping.

4.6 GIS PROGRAMS FOR MAPPING

Geographic information systems (GIS) are used to input, analyze, manipulate and display geographic information. The technology developed rapidly in the 1980s and has grown into a multibillion dollar business. Typical applications include map overlay, point in polygon analysis, buffer analysis, and address matching. GIS are used by government agencies, private business, and research institutions. The requirements of the data base usually necessitate larger computers.

ARC/INFO® (Environmental Systems Research Institute) stores location data with both coordinates and topology. The topological data are used to identify spatial relationships between arcs, nodes, and polygons. The program can produce different types of maps, but the display of qualitative and quantitative data with the choropleth technique is the major form of output.

Other GIS programs have been specifically designed for the microcomputer. Atlas GIS (Strategic Mapping, Inc.) for Windows performs most functions typical of a GIS. The map data is stored in a parcel or polygon data structure that limits some of the overlay and address-matching capabilities. The program creates choropleth, dot, cartogram, point-symbol, and two-variable maps. Another example of a GIS program for the microcomputer is MapGraphix (Comgraphix) for the Macintosh. This program includes the standard functions of digitizing, editing, spatial registration, pattern fills, and custom symbols. The program supports data base links to programs such as FoxBase (Microsoft) and 4th Dimension (Acius). Map output consists of choropleth maps and maps with user-defined symbols. The program also stores map data in a parcel/polygon data structure.

Raster-based GIS programs, including ARC/INFO® GRID (ESRI), GRASS (USACERL), IDRISI (Clark University), and MAP II (ThinkSpace), incorporate extensive mapping functions. These programs store geographic data in separate layers and have a variety of functions to manipulate and compare layers. A mapping capability that is shared by many of these programs is the ability to "drape" a layer of data (e.g., vegetation or a digital image) over a three-dimensional relief map.

4.7 SUMMARY

Computer mapping was made possible by the development of a variety of computer graphics hardware devices. The vector-based peripherals of the 1970s, including the coordinate plotter and the storage tube, gave way a decade later to raster-based devices such as the raster-graphics terminal and laser printer. But the purpose of computer mapping remained to produce a map on paper. Graphics terminals, for example, were mainly used to preview a map before it was sent to a plotter or printer.

A number of highly interactive computer mapping programs have been developed for the microcomputer. These can be categorized into three major classes: (1) statistical analysis programs, (2) desktop mapping, and (3) programs for Geographic Information Systems. In terms of the number of programs developed desktop mapping has been the most dynamic category.

4.8 EXERCISES

1. Use a desktop mapping program to create a choropleth map. Display the data with different classification options.

2. Use a contour mapping program such as SURFACE III or SURFER to map the high temperatures for 20 selected cities on a certain day.

3. Use a map projection program to create a perspective projection of the earth that is centered on your location.

4. Use a digitizing program such as ROOTS to digitize a map with at least 50 polygons.

5. The following listing is a SAS/GRAPH file to create a map of the United States depicting infant mortality in 1982. Note that the GOPTIONS and LIBNAME commands need to be changed depending on your system configuration. GOPTIONS specifies the graphic output device and LIBNAME indicates the location of the map files.

Type the commands below into a file. Do not type the comments that appear in parenthesis. Submit the file for execution by typing SAS <filename>. If a map is not drawn on the screen, look at the file called <filename>.LOG.

Use a terminal emulator such as Telnet. Telnet for the PC does not emulate a graphics terminal. If you are using a PC, you will need to use another terminal emulator that supports graphics.

(The first line indicates the graphics output device. This line will change depending upon the graphics terminal or terminal emulator you are using.)

```
GOPTIONS DEVICE=TEK4014;
```

(The following line specifies the location of the SAS/GRAPH map files on your computer system. This line may change depending upon how the program was installed. If this does not work, ask your system manager where the SAS/GRAPH maps are stored.)

```
LIBNAME HERE 'DKA0:[SAS607.MAPS]';
DATA STATES;
   SET HERE.US;
DATA VAL;
    INPUT STATE INFMORT;
CARDS;
```

(The first column is the FIPS code and the second is the Infant Mortality rate for 1982 in number of deaths per 1,000 live births. Do not type the third column – the name of the state. This is included here for informational purposes only.)

```
 1   13.80          Alabama
 4    9.30          Arizona
 5   10.10          Arkansas
 6    9.89          California
 8    9.10          Colorado
 9   11.10          Connecticut
10   14.10          Delaware
12   12.80          Florida
13   12.69          Georgia
16    9.89          Idaho
17   13.60          Illinois
18   11.39          Indiana
19   10.19          Iowa
```

20	10.39	Kansas
21	12.00	Kentucky
22	13.00	Louisiana
23	9.00	Maine
24	11.89	Maryland
25	10.10	Massachusetts
26	12.10	Michigan
27	9.50	Minnesota
28	15.39	Mississippi
29	11.69	Missouri
30	10.10	Montana
31	10.00	Nebraska
32	10.19	Nevada
33	11.00	New Hampshire
34	11.69	New Jersey
35	11.30	New Mexico
36	12.10	New York
37	13.69	North Carolina
38	10.60	North Dakota
39	11.50	Ohio
40	12.30	Oklahoma
41	10.50	Oregon
42	11.60	Pennsylvania
44	10.00	Rhode Island
45	16.10	South Carolina
46	10.19	South Dakota
47	12.00	Tennessee
48	10.89	Texas
49	11.00	Utah
50	9.30	Vermont
51	12.80	Virginia
53	10.60	Washington
54	11.39	West Virginia
55	9.50	Wisconsin
56	9.80	Wyoming

```
PROC GMAP DATA=VAL MAP=STATES;
    ID STATE STATE;
```
(The following statement specifies the type of map. Options are CHORO, SURFACE, PRISM, BLOCK.)

```
    CHORO INFMORT / LEVELS=5 CTEXT=WHITE COUTLINE=WHITE;
```

(The following PATTERN statements define the shading patterns for the five categories. "V" specifies a line shading. "M2" specifies the level of the shading with "M3" darker than "M2". "N" indicates single lines. "X" is for a cross-hatched shading. "E" stands for empty. "S" is for solid.)

```
      PATTERN1 V=E ;
      PATTERN2 V=M2N45 C=WHITE;
      PATTERN3 V=M3X45 C=WHITE;
      PATTERN4 V=M4X45 C=WHITE;
      PATTERN5 V=S C=WHITE;
      TITLE1 H=2 F=XSWISSE C=WHITE 'Infant Mortality — 1982';
RUN;
```

4.9 REFERENCES

BRASSEL, K. (1974) "A Model for Automated Hill Shading." *The American Cartographer* 1: 15–27.

CERNY, J. W. (1972) "Use of the SYMAP Computer Mapping Program." *Journal of Geography* 71, no. 3: 167–174.

COOK, P. G. (1980) "Graphic Display Technology and Applications." In *Computer Graphics Hardware: Harvard Library of Computer Graphics*. Cambridge, MA: Harvard University Press, 41–45.

DOUGLAS, D. H. AND PEUCKER, T. K. (1973) "Algorithms for the Reduction of the Number of Points Required to Represent a Digitized Line or Its Caricature." *Canadian Cartographer* 10, no. 3: 110–122.

FREIBERGER, P. AND SWAINE, M. (1984) *Fire in the Valley: The Making of the Personal Computer*. Berkeley, CA: Osborne/McGraw–Hill.

GOODCHILD, M. F. (1988) "Stepping Over the Line: Technological Constraints and the New Cartography." *The American Cartographer* 15, no. 3: 311–319.

LIEBENBERG, E. (1976) "SYMAP: Its Uses and Abuses." *Cartographic Journal* 13, no. 1: 26–36.

MACHOVER, C. (1980) "Graphic CRT Display State-of-the-Art." In *Computer Graphics Hardware: Harvard Library of Computer Graphics*. Cambridge, MA: Harvard University Press, 61–63.

MONMONIER, M. S. (1982) *Computer-Assisted Cartography: Principles and Prospects*. Englewood Cliffs, NJ: Prentice Hall.

SCHWARTZ, K. D. (1993) Embedded Map Technology Beats a Path to the Desktop. *Government Computer News* 12, no. 7: 47.

SHEPARD, D. (1968) "A Two-Dimensional Interpolation Function for Irregularly-Spaced Data." *Proceedings, Association for Computing Machinery*, 23rd National Conference, 517–524.

SHURKIN, J. (1984) *Engines of the Mind: A History of the Computer*. New York: W. W. Norton.

FURTHER READINGS:

ARONOFF, S. (1989) *Geographic Information Systems: A Management Perspective.* Ottawa, Ontario: WDL Publications.

BURROUGH, P. (1986) *Principles of Geographical Information Systems for Land Resources Assessment.* Oxford: Clarendon Press.

SALMON, R. AND SLATER, M. (1987) *Computer Graphics: Systems and Concepts,* Reading, MA: Addison-Wesley.

TOMLIN, D. (1990) *Geographic Information Systems and Cartographic Modeling,* Englewood, NJ: Prentice Hall.

5

Graphic Processing

5.1 INTRODUCTION

Graphic processing is the interactive creation of graphic illustrations with the computer. It may be viewed as the graphic counterpart to text processing. Graphic processing developed as the direct result of graphic input devices such as the mouse and programs that supported interactive graphic editing. Programs for graphic processing have had a major influence on the production of maps.

In many ways graphic processing was a greater advance than text processing. Before the computer the typewriter could be used to bring the written word to paper. However, no machine existed that made the creation of a graphic easier. There was no counterpart to the typewriter to assist in the creation of a graphic illustration. It was done by hand with a drafting pen, a straight edge, and a variety of other tools. Special skills and training were necessary to work with these tools. Because of the difficulty in creating an illustration by hand, text was often used to express a concept for which a graphic would have been much more effective. Not only is it now possible to create illustrations with the computer, they can also be easily incorporated within text documents. Graphic illustration programs have changed the way information is communicated.

The importance of graphic illustration programs to cartography is twofold: (1) the programs can be used to interactively create maps in a process that is analogous to hand-drafting, and (2) the programs can edit the maps produced by computer mapping and GIS programs. Both aspects represent another form of interactivity in cartography — interactivity in the creation and editing of maps.

Programs for graphic processing can be grouped into two major categories: (1) Paint programs use a raster data structure and make the pixel-by-pixel editing of an illustration possible, and (2) object-based programs use a vector-type data structure and represent a graphic as a series of individual objects. In this chapter we examine the graphic processing concept, how it has been implemented in a variety of programs, and how it is used in cartography.

5.2 PAINT GRAPHICS

Underlying the paint graphics program is a raster or grid of pixels. Each pixel is treated independently and can be edited "dot by dot." Different names are given to the grid depending on the number of gray shades or colors that can be represented. A bit-map is a grid that consists of black or white pixels each represented by 1 bit (8 bits equal a byte). A byte-map is a grid of gray or color pixels each represented by 1 byte (8 bits). Up to 256 levels of gray or color may be represented with this number of bits. To represent more colors, more bits per pixel are needed. If one byte is assigned to each of the three primary additive colors — red, green, and blue — then up to 16,777,216 (2^{24}) colors can be represented.

Paint programs not only share a common raster data structure but also a similar set of drawing tools. These tools are usually grouped together in a graphic menu palette that includes lines, arcs, rectangles, circles, polygons, letters, and other graphic elements. The tools are used to change the current state of individual pixels within the matrix of pixels. Additional tools are available to fill an enclosed shape with a shading, move or erase parts of the display, and zoom into the graphic to edit single pixels.

The major limitation of paint graphics is in the enlargement or reduction of the graphic. Enlargements based on whole numbers (e.g., 2x, 3x, 4x) are fairly easy to accomplish. Each pixel is simply duplicated by the square of the enlargement factor (Figure 5.1). A 2x enlargement requires that four pixels be made from each pixel, 3x leads to nine pixels. Reduction in size can be similarly accomplished but results in a loss of information. To reduce a graphic 2x, one of the four original pixels must be selected, or the average of the four found. Enlargements or reductions that are not based on a whole number are more difficult to accomplish, and the results are not as satisfactory. This problem with paint graphics is especially apparent with printers such as the laserwriter, that have a higher resolution than the screen.

Paint programs have many advantages. The function of different menu items and tools can be learned by experimentation. A zoom function makes it possible to enlarge and edit portions of the display selectively in a pixel-by-pixel fashion. The raster data structure of paint graphics is directly compatible with video memory and can be displayed quickly on the screen. This is useful when the graphic is to be displayed as part of an animation. In fact, programs for computer animation often incorporate paint graphic editors.

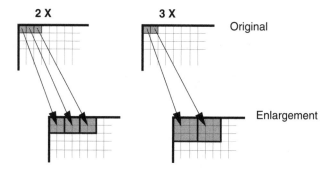

Figure 5.1   A raster zoom. The enlargement or reduction of a raster graphic is easily performed when the amount of reduction or enlargement is represented by a whole number (2, 3, 4, ...). In the case of a zoom each pixel is duplicated by the square of the enlargement factor (4, 9, 16, ...).

5.2.1 Programs for Paint Graphics

The program that first popularized paint graphics was MacPaint. Introduced with the Apple Macintosh, the program was among the first to incorporate a graphical menu bar. Like the original Macintosh, the program was initially limited to black and white illustrations. Many other paint programs are now available and most support color.

Canvas (Deneba) and SuperPaint (Adobe) are widely used programs for paint graphics, both supporting paint and object graphics. The function of each item within the graphical menu palettes of these programs is listed in Figure 5.2. The tools can be categorized as: (1) selection tools that choose a part of the graphic, (2) drawing tools that draw an object, and (3) options tools that change a characteristic of an object such as line thickness. A comparison of the two palettes indicates some standardization in icons for individual tools.

A number of other microcomputer programs have essentially the same functions and operate in the same general way (e.g., Microsoft Paint for Windows, Windows Paintbrush (Microsoft), Dr. Halo (Media Cybernetics)). The icons used in the menu palettes are often identical. The output from these programs can also be incorporated within text processing documents.

There are two other types of raster graphics programs. The first, color paint programs, are the color counterpart to bit-map paint programs. They have the same graphic functions but work in color and are specifically designed for use by artists (e.g., Claris BrushStrokes, Painter from Fractal Design). The second type, image processing programs (described in the next chapter), may have similar pixel-based editing functions but can also process the image as a whole (e.g., Adobe Photoshop for Macintosh and Windows). These programs are mainly used for the editing of digitized color

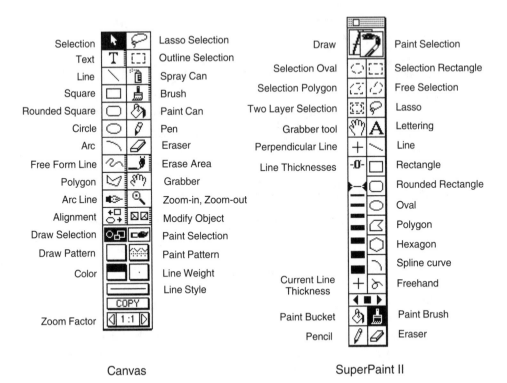

Selection		Lasso Selection
Text		Outline Selection
Line		Spray Can
Square		Brush
Rounded Square		Paint Can
Circle		Pen
Arc		Eraser
Free Form Line		Erase Area
Polygon		Grabber
Arc Line		Zoom-in, Zoom-out
Alignment		Modify Object
Draw Selection		Paint Selection
Draw Pattern		Paint Pattern
Color		Line Weight
		Line Style
Zoom Factor		

Canvas

Draw		Paint Selection
Selection Oval		Selection Rectangle
Selection Polygon		Free Selection
Two Layer Selection		Lasso
Grabber tool		Lettering
Perpendicular Line		Line
Line Thicknesses		Rectangle
		Rounded Rectangle
		Oval
		Polygon
		Hexagon
		Spline curve
Current Line Thickness		Freehand
Paint Bucket		Paint Brush
Pencil		Eraser

SuperPaint II

Figure 5.2 Canvas and SuperPaint graphical menu palettes. The two palettes use many of the same icons to indicate the drawing options. (Courtesy of Deneba Software and Adobe Corp.)

photographs and come with an array of filters for sharpening, blurring, softening, adding motion and distorting images. Composite images can also be pieced together from several pictures.

5.2.2 Maps with Paint Graphics

The quality of printed output from paint programs is usually no better than the resolution of the screen, generally between 70 to 100 dpi. Cartographers, concerned with graphic quality on paper, generally prefer to produce maps at the resolution of the printer (between 300 and 3386 dpi). However, if maps are being used directly from the screen, there is no difference in graphic quality between the paint and object approaches. The raster approach is useful for an animated display of maps because the images can be placed on the screen directly without having to be drawn.

The starting point for creating maps with a paint program is some type of base map. A base maps can be drawn with a digitizing pad or input with a scanner. Both options can be time consuming. Maps that are scanned usually need to be "cleaned" by reducing lines to a consistent thickness. Inputting the maps with a digitizing pad is very difficult because the programs use a relative positioning system. An alternative is the use of existing paint graphic maps, representing a type of software sometimes referred to as "clip-art." Clip-art map files are available from several sources (e.g., Cartesia Software Inc.).

Shaded relief maps are one type of base map available in the raster format (Figure 5.3). The shaded relief method assumes a light direction from the northwest. Slopes that

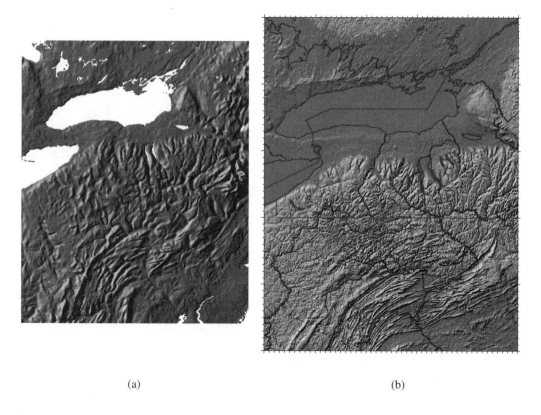

(a)   (b)

Figure 5.3   Shaded relief maps of the northeastern United States and southern Canada. Both are distributed in a raster format and can be edited with color paint and image processing programs.   Image (a) is from Digital Wisdom, Inc. (see Appendix B).   Image (b) is from the ftp site "fermi.jhuapl.edu" (see Appendix C.)

face that direction are illuminated whereas slopes that face the opposite direction — southeast — are shaded. The technique provides a realistic depiction of elevation.

Shaded relief maps in digital form can be edited with color paint and image processing programs. Editing can be done pixel-by-pixel or based on channels. A channel divides the image into layers. A color image has four separate channels — red, green and blue (RGB, the additive primary colors), and a channel to show the combined colors. A color image using the subtractive colors also has four channels — cyan, magenta, yellow, and black (abbreviated as CMYK). Additional channels can be added to an image, and these can be used to store selections or masks that are used to make selections. For example, a selection could be used to separate land from water. Channels are useful in editing specific parts of the image.

5.2.3 Paint Graphic Maps

The following maps illustrate the use of a paint graphics program for the production of thematic maps. The icons presented with each map represent the tools that were used to draw each map. Figure 5.4 illustrates two types of point-symbol maps created with a paint graphics program.

Figure 5.5 depicts a graduated bar map. The bars are placed on the map after being created separately. Before filling the bars with a shading, the "Trace Edges" menu item was used to add an outline.

Figure 5.4   Two types of point-symbol maps created with a paint program. The icons below each map indicate the tools that were used to create the map. The state outlines were derived from an existing map of the United States in paint format.

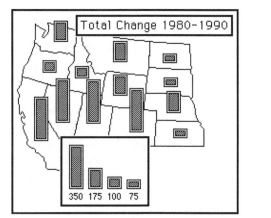

Figure 5.5   Bar map created with a paint program. The four different-sized bars were drawn using a ruler function. Each was then duplicated and moved onto the map.

For the choropleth map in Figure 5.6 the "Trace Edges" option was used to outline the states. Tracing the edges has the effect of introducing a white line between black polygons. The paint can tool was then used to fill each area with a shading.

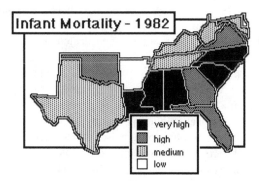

Figure 5.6   Choropleth map created with a paint program. All of the line work in the map was first traced with the "Trace Edges" function. This has the effect of converting a simple line to a double line. Polygons that are later filled with a black shading will then be separated by a white line.

5.3 OBJECT GRAPHICS

Object graphics is based on individual, editable graphic objects such as lines, rectangles, circles, and polygons. The objects are defined internally as either a list of x, y coordinates or a geometric function, such as a Bezier curve. The precise coding of each object depends upon its form. For example, lines and polygons are defined with a list of x, y coordinates; a rectangle with two x, y coordinate pairs, one for the upper-left and one for the lower-right corners; circles are defined with a middle x, y coordinate and a radius value; and a curved line is represented with a Bezier formula that characterizes the line. Objects can also be assigned characteristics such as line thickness or line color, and they can be filled with a pattern or color. Although the objects are displayed on the screen as pixels, the underlying geometric definition allows them to be selected and modified much more easily than with paint graphics.

The object approach to graphics overcomes many of the limitations of paint-type graphics. The major advantage is the ease with which the graphic can be scaled. Scaling of a graphic object is accomplished by selecting an object and either expanding or reducing a bounding rectangle by clicking and dragging on its graphics *handles* (small black rectangles that surround a selected object). Some programs allow an enlargement or reduction factor to be entered. The factor can be either less than one for a reduction or more than one for enlargement, and can be different for horizontal and vertical dimensions. The ability to easily scale the graphic makes it possible to print an illustration at a higher resolution than is visible on the screen.

Another advantage of the object approach is that the graphic is more editable. Individual objects can be selected, moved, enlarged, reduced, and rotated. In paint graphics once a figure is filled with a shading, the shading cannot be changed without erasing all of the pixels within the figure. In object graphics changing the shading of a figure is done by simply defining a different shading for a selected object.

Object graphics programs are used extensively for the creation of maps. The programs are used in two ways: (1) creating a map interactively, most often beginning with a base map, and (2) editing the output of a computer mapping or program. Most maps that are now published in books, newspapers, and magazines, including most of the maps in this book, are created with object graphics programs.

There are three types of object graphics programs: *CAD*, *Draw*, and *Illustration*. Computer-aided design or CAD programs, sometimes called computer-aided drafting and design (CADD), have been available since the 1970s. The programs essentially automate the traditional drafting process. *Draw* programs incorporate a subset of CAD tools for more general graphic illustration purposes. *Illustration* programs are intended primarily for high-resolution printing applications.

5.3.1 CAD Programs

Computer-aided design (CAD) is the computer equivalent of mechanical drafting. CAD programs are also used in cartography, especially for facilities mapping applications (e.g., location of water and sewer lines). CAD programs were originally written for minicomputers and a major market developed in the 1970s for turnkey CAD systems that combined specialized graphics hardware and software. By the early 1980s companies such as Intergraph were purchasing large numbers of minicomputers that were resold as part of integrated CAD systems. High-end CAD applications have migrated to computer workstations, and these still represent a large portion of the CAD market. In the case of Intergraph, workstations have been designed specifically for graphics applications. By number of users, however, the most popular CAD programs today are for microcomputers.

A distinguishing characteristic of CAD programs is the use of layers. Each layer contains only part of the overall graphic. For mapping applications one layer might represent streets, a second layer would be the electric network, a third layer would be the water lines. Each layer can be viewed and edited individually. The layer approach is conceptually simple but can limit certain types of analysis. In addition, it is rarely possible to view all of the layers at once because features will be drawn on top of each other.

There are many different CAD programs for microcomputers. MiniCAD for the Macintosh is typical of these programs (Figure 5.7). As with paint graphics programs, graphic tools are arranged in palettes on one side of the screen. In addition to the line, square, and circle tools, CAD programs also have graphic tools for constrained lines (drawn at 0 or 90 degrees) and quarter and full arcs. The programs emphasize positional accuracy and usually display the current mouse position in inches or centimeters. Objects can be scaled, rotated, and stretched.

Probably the most widely used CAD program is AutoCAD, available for the PC, Macintosh, and UNIX workstations. AutoCAD includes three-dimensional wireframe modeling, surface modeling, and a multiple viewport facility that provides multiple views of a drawing from different perspectives. Other programs for the PC include EASYCAD and FASTCAD (Evolution Computing) and TurboCad Designer (Imsi).

Maps for CAD

The industry standard Drawing Interchange File (DXF) format from AutoCAD is commonly used for the distribution and exchange of map files in a CAD format. A variety of maps can be purchased in this format, including maps of the United States that were originally produced by the United States Geological Survey and the U.S. Census Bureau, maps of zip code areas in the United States and maps of the world (e.g., Micro Map & Cad).

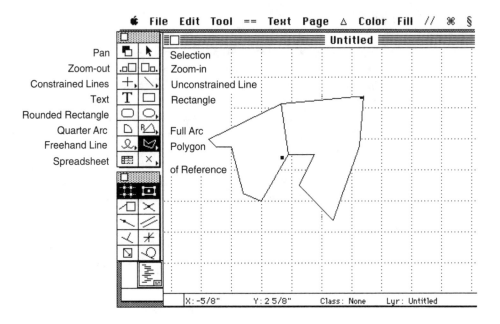

Figure 5.7   Menu palette and drawing area of MiniCAD. CAD programs also make use of a graphic menu palette that incorporates the various drawing options.

5.3.2 Draw Programs

Draw programs, incorporating only a subset of CAD functions, are designed for the sketching and general graphic illustration work. Although programs in this category are used primarily for the creation of more general graphic illustrations, they are also used for technical illustrations and architectural design. The programs have a more intuitive user interface and are easier to learn than most CAD or illustration programs.

Draw programs include standard tools for placing text and the drawing of geometric objects. Objects can also be filled with a shading, rotated, or resized. Although these programs can output to higher-resolution PostScript printers, they generally do not take complete advantage of their functionality or resolution as do illustration programs. Therefore, Draw programs are not used as often for the creation of maps.

MacDraw (Claris) was one of the first Draw programs and has been widely imitated. MacDraw's graphic menu palette and Arrange menu include its basic functions (Figure 5.8). The selection tool is used to select objects or outline a group of objects that have already been drawn. The eight functions below the selection tool are for drawing individual graphic objects. The bottom item in the menu is for changing the scale, and the number in the menu item above it indicates the current magnification (from 12.5 to 3,200 or $\frac{1}{8}$ of the original to 32 times as large; 100 indicates the illustration is at a 1:1

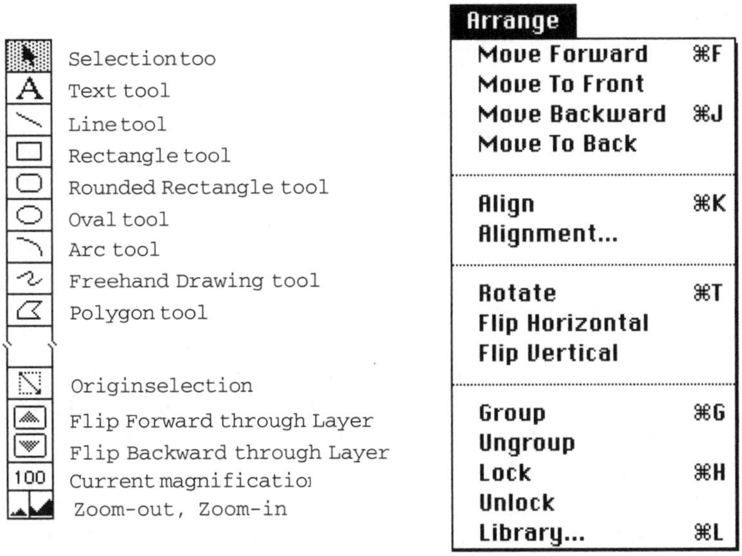

Figure 5.8   Graphic menu palette and Arrange menu from the MacDraw program. The palette represents the possible drawing functions. The Arrange menu allows changes to be made to the objects, including positioning objects on top of or behind other objects, aligning objects in a particular arrangement, rotating objects, grouping and ungrouping and locking objects so that they cannot be accidentally altered. (Courtesy of Claris Corp.)

scale). As with CAD programs, MacDraw can create an illustration with multiple layers. The two menu items above those for scale selection make it possible to flip through the individual layers. *Origin selection* switches the beginning point of an object between either the middle or the corner of the object.

The MacDraw *Arrange* menu is for the manipulation of graphic objects (Figure 5.8). The first four items control the visibility of objects relative to other objects by assigning a *depth characteristic*. This facilitates the placement of objects in front of or behind each other. The next items deal with the alignment of objects and can be used, for example, to center a series of objects on the middle axis of a page. The *Rotate* and the *Flip* functions change the orientation of graphic objects. The *Group* function allows several objects to be defined as a single object. *Ungroup* separates this grouping. *Lock* freezes a graphic object in place, and *Unlock* allows it to be moved and manipulated again. *Library* stores frequently used graphic objects by name.

Other draw programs include DrawPerfect (WordPerfect), and Windows Draw (Micrografx). DrawPerfect is a graphics presentation program for PCs that works in conjunction with WordPerfect, a popular text processing program, and is therefore the most widely distributed. It is designed for producing illustrations for slides and text documents. The program incorporates a graphic palette, pull-down menus, and scroll bars. The program also supports the creation of graphs using imported spreadsheet data.

Maps with Draw Graphics

As discussed, there are many advantages to the object approach to graphics. Moving and resizing are accomplished more easily, and the printed maps have a higher resolution. However, creating the object base maps is even more difficult than with paint graphics. The automatic conversion of scanned images to objects through an *Autotracing* procedure does not work properly for maps (see following section). "Screen-digitizing" using a bit-mapped graphic as a background is a tedious process. One solution is to use a digitizing program such as Roots to input the map. Completed base maps are also available for selected areas (e.g., MapArt from Cartesia Software).

Figures 5.9 and 5.10 illustrate maps created with MacDraw. The bar map was created by subsetting the desired states from an object formatted map file of the United States. Bars were added with the rectangle tool. The choropleth map in Figure 5.10 was constructed by filling each state with the proper shading and adding the legend and text objects. It is not possible to specify white lines with MacDraw as can be done with some programs; therefore, the states filled with the black shading are not separated by a white line.

Adding a three-dimensional effect to a map is possible by overlaying two slightly offset maps. The drop shadow effect is commonly applied to maps that appear in newspapers and magazines. The procedure is accomplished by offsetting a duplicated map object, filling it with black or a dark shading, and moving it to the back, behind the original map (Figure 5.11).

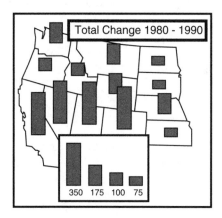

1. Select and subset polygons of western states from map of the US.

2. Choose a medium grey shading and create four bars with heights proportional to the values depicted.

3. Duplicate individual bars and move on top of states.

4. Add title and legend numbers.

5. Change line width to 2 pts. and draw legend and neatline boxes.

Figure 5.9   Bar map created with MacDraw. The procedure is similar to the creation of a map with a paint graphic program, except that objects can be more easily moved and manipulated.

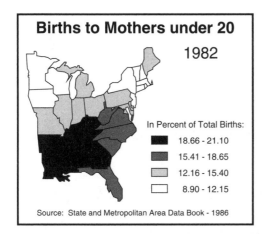

Births to Mothers under 20

1982

In Percent of Total Births:

18.66 - 21.10

15.41 - 18.65

12.16 - 15.40

8.90 - 12.15

Source: State and Metropolitan Area Data Book - 1986

1. Select and subset polygons of eastern states from a map of the US.

2. Select individual states and fill with the proper shading.

3. Create legend by drawing a single rectangle with the rectangle tool and duplicating it three times. By moving the second rectangle to its proper desired position below the first, subsequent duplicated rectangles will be placed in the correct position. Shade each rectangle with the corresponding shading.

4. Add text fields and neatline.

Figure 5.10   Choropleth map created with MacDraw. Polygons are assigned a shading value. Note that the polygons shaded with a black shading are not separated by a white line. Some programs allow the lines separating these polygons to be drawn in white.

Step 1. Duplicate a map object and offset to the bottom right.

Step 2. Fill the offset map with black.

Step 3. Move the offset map behind the original map.

Figure 5.11   Creating the drop shadow effect. The procedure consists of duplicating the map, filling it with black, and then moving it behind the original map. The "Bring to Front" and "Send to Back" functions in the Arrange menu are used to layer objects on top of each other.

5.3.3 Illustration Software

Illustration programs are based on the PostScript page description language (PDL) that has become a standard way of communicating to laser printers and imagesetters. PostScript, developed by Adobe, is a computer language for the display of text and graphics. The language incorporates a wide range of graphic operators and allows the combining of operators into more complex procedures and functions. PostScript printers contain programs to interpret PostScript commands and rasterize the graphic for printing. They have resolutions between 300 and 3,386 dots per inch.

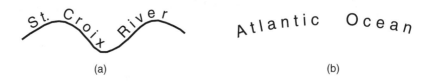

(a)   (b)

Figure 5.12   Flowing text along curve with Postscript object graphics. In (a) the text is offset from the line by 4 points. In (b) the text is attached to a curved line but the line is not shown.

Illustration programs share a number of common characteristics. The programs emphasize both graphic illustration and text handling. Text can be placed along a curve (Figure 5.12) or flowed within the boundaries of graphic objects. Drawing options usually include the ability to trace over scanned images manually or automatically, editing points within object boundaries, and the ability to blend two objects with a continuum of colors, gray shades, shapes, and line weights. Transformation options include the rotation of objects, the scaling of objects uniformly or nonuniformly, the reflection of objects along different axes, and the shearing of objects. The programs also support spot, tint, or process colors (CMYK and RGB), including PANTONE© colors (a standard color set) and the creation of color separations. An autotrace function for the conversion of a bit-map graphic is a standard feature of these programs.

One of the first illustration programs was Adobe Illustrator (available for the Macintosh, PCs and UNIX workstations). The program combines text handling, illustration, and graphing capabilities for single-page design and layout. Illustrator's graphic menu bar and Arrange menu is similar to that of MacDraw (Figure 5.13). The first eight items in Illustrator's graphic palette can also be found in MacDraw, although some use different icons. The nine menu items that follow the circle tool are indicative of some of the additional functions available in illustration programs. *Gradation fill* blends two objects into one with either a series of shadings, colors, or shapes. The scaling tool scales objects to different sizes either with the mouse or by entering a magnification or reduction factor. The *Rotate*, *Reflect* and *Shear* tools transform an object. The *Cut* tool introduces a new point within a line. The *Measure* tool is used to determine the size of an object. The *Position Page* tool places the page outline within the working area of the page, and the *Graph* tool accesses the bar, line, area, scatter, and pie graph options.

Illustrator's Arrange menu incorporates the typical object-graphics functions, including *Group* and *Lock*. *Transform Again* duplicates the action of the last graphic transformation. *Join* connects two lines, and *Average* finds the average position of all points within an object. *Hide* and *Show All* control the visibility of objects. Guides are used as a drawing aid, and crop marks delimit parts of the graphic display.

Other programs in this category include FreeHand from Altsys Corporation

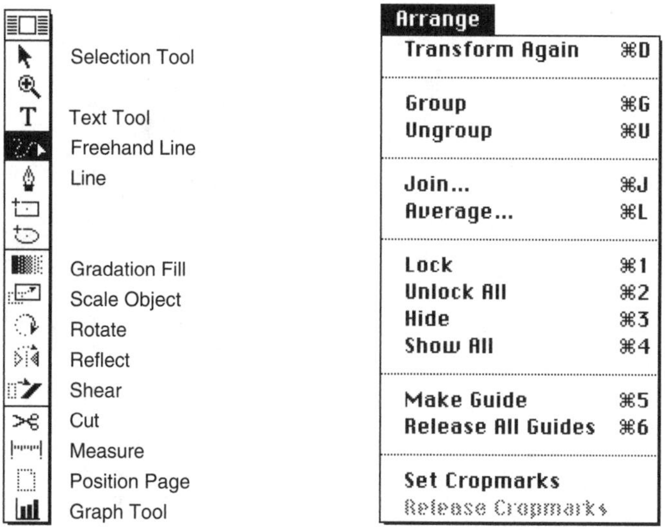

	Selection Tool	
	Text Tool	
	Freehand Line	
	Line	
	Gradation Fill	
	Scale Object	
	Rotate	
	Reflect	
	Shear	
	Cut	
	Measure	
	Position Page	
	Graph Tool	

Arrange

Transform Again	⌘D
Group	⌘G
Ungroup	⌘U
Join...	⌘J
Average...	⌘L
Lock	⌘1
Unlock All	⌘2
Hide	⌘3
Show All	⌘4
Make Guide	⌘5
Release All Guides	⌘6
Set Cropmarks	
Release Cropmarks	

Figure 5.13   Graphic palette and Arrange menu from Adobe
Illustrator. The palette and menu have many of the same options as
MacDraw. Courtesy of Adobe Systems, Inc.)

(Macintosh and Windows), CorelDraw from Corel Corp. (DOS, Windows, and UNIX)
and Designer from Micrografx (DOS and Windows). FreeHand supports pressure-
sensitive tools for the control of line thickness. CorelDraw incorporates paint graphic
functions in addition to object functions. Designer supports the blending of two objects
though a metamorphosis procedure.

Maps in Print

Illustration programs are commonly used for the production of the graphic
illustrations that are published in newspapers, magazines, and books (including this one).
Newspapers have become one of the major users of graphic illustration software. The
graphics departments in most newspapers have converted to illustration software.

GraphicsNet, begun in October 1991, is a wire service of the Associated Press that
provides newspapers with graphical illustrations and maps. The files are transferred over
a network to computers at local newspapers on a continuous basis. These files reside on
the local computer for approximately three days and then are automatically purged.
During this time the newspaper may select any of the illustrations that are provided.
They may also edit the illustration before inclusion in their newspaper. All of the files
that are available through GraphicsNet are in a format compatible with illustration
software.

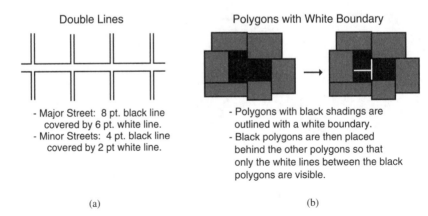

<table>
<tr><td>Double Lines</td><td>Polygons with White Boundary</td></tr>
</table>

- Major Street: 8 pt. black line
  covered by 6 pt. white line.
- Minor Streets: 4 pt. black line
  covered by 2 pt white line.

- Polygons with black shadings are
  outlined with a white boundary.
- Black polygons are then placed
  behind the other polygons so that
  only the white lines between the black
  polygons are visible.

(a)

(b)

Figure 5.14   A variety of graphic tricks are possible with illustration programs. For example, a double line can be created by placing a thinner white line on top of a thicker black line. It is also possible to make the boundaries of black polygons visible by specifying that their outline be white.

Graphic Tricks with Illustration Programs

A number of graphic tricks have developed for use with illustration programs. These tricks consist of a particular combination of operations that achieve a certain effect. Many of these tricks are useful in the production of maps. For example, a road represented with a double line is created by placing a thinner white line on top of a thicker black line, as demonstrated in Figure 5.14a with the major and side streets. In this case the white interior lines of the side streets are on top of the major street to create a continuous white space. In Figure 5.14b the interior polygons are outlined with a white line to make their boundaries visible. These polygons are placed behind the other polygons so that only the white lines separating the black polygons remain visible.

5.4 PAINT TO OBJECT GRAPHICS CONVERSION

All Illustration programs have an Autotrace function that converts a paint graphic into an object graphic. This raster-to-vector conversion process involves a line following procedure through a grid or bit-map. The function essentially converts a selected portion of the paint graphic to an object format. This can be done on an object-by-object basis or by converting the entire graphic at once. Autotracing is the main method of converting scanned images to object graphics.

The *Autotrace* function generally works well with single polygons (Figure 5.15). Most tracing procedures have difficulty determining the individual polygons if the paint graphic includes multiple areas with shared boundaries. Figure 5.16 is the result of the autotracing of a bit-map representation of multiple states. Only the state of Nevada has

Paint   Object
                                   (after AutoTrace)

Figure 5.15   Paint and autotraced object versions of a polygon. The Paint polygon is represented with black pixels against a white background. The Object polygon is represented with a series of x, y coordinates.

been traced correctly because it is the only interior polygon. The tracing procedure does not convert all of the polygons and uses a large number of points to represent the very simple polygons that are converted. A total of 89 points are used to outline the state of Nevada when approximately 7 points would have been sufficient at this scale. The excessive number of points results from the unnecessary stair-step encoding of diagonal lines.

Illustration programs that display a bit-map graphic as a background figure provide tools to interactively control the autotrace process. With Illustrator, for example, polygons can be converted individually (Figure 5.17). This is accomplished with an autotrace tool that is placed near a bit-map line or polygon. This procedure works well when the bit-map contains "clean" lines. Usually, however, the scanning process will introduce extra pixels or noise with the line work. The more tedious *screen digitizing* process must then be used to manually draw over the scanned image. The bit-map graphic can be used as a guide for this manual tracing process.

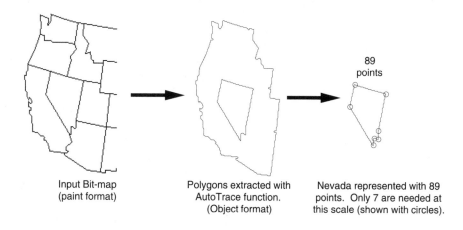

89
points

Input Bit-map
(paint format)

Polygons extracted with
AutoTrace function.
(Object format)

Nevada represented with 89
points. Only 7 are needed at
this scale (shown with circles).

Figure 5.16   Example of the autotracing of multiple polygons. The input bit-map has not been traced properly. Only Nevada, is correctly traced, and it is represented with more points (89) than are necessary at this scale.

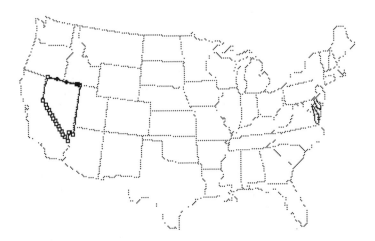

Figure 5.17 Conversion of individual polygons from a bit-map graphic into objects with Illustrator. States on the east coast cannot be distinguished from each other in the bit-map graphic. Also notice that the diagonal line in Nevada is represented with more points than are needed to represent this straight line.

5.5 MAPS WITH POSTSCRIPT

The PostScript page description language provides a number of advanced features, including a method for defining a continuum of dot shadings between white and black. With Illustration programs gray shadings can be defined for each polygon as a percentage ink value. This feature makes it possible to create a new type of choropleth map.

In 1973 Waldo Tobler introduced a computer procedure that shaded each individual area in proportion to its specific value, thereby eliminating the need to classify data. This ungeneralized representation of quantitative area distributions is referred to as the unclassed method of choropleth mapping. The unclassed method of mapping was criticized by some for presenting too much information (Dobson 1973), but after testing it was found that such maps could be interpreted at least as well as classified choropleth maps (Peterson 1979).

One problem associated with any symbolization is the possible discrepancy between perceived and actual values. This phenomenon can best be understood by observing that an area covered with 50 percent ink is *not* perceived as halfway between white and black. In general, humans underestimate value, defined in percent reflectance, so that a shading with greater than 50 percent white (less than 50 percent black) is perceived as the middle shading.

In Illustrator, shadings are assigned to polygons with the *Style* menu item. Figure 5.18 depicts a series of shadings created with Adobe Illustrator. The shadings

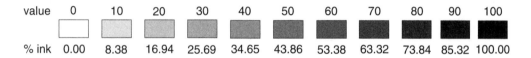

Figure 5.18   A ten-shading continuum created with Illustrator. In general, humans underestimate value, where value is defined as percent reflectance. To compensate for this underestimation, reflectance values are increased by decreasing the amount of ink.

have been perceptually adjusted using the formula by Williamson (1982):

$$W = P^{.8333} / 0.464$$

where, W is the gray tone in percentage of area white and P is the desired perceived value. A desired value of 50 (P) equals a 56.14 percent white or 43.86 percent ink.

Illustrator files are in PostScript text format and can be edited with a text processor. The files consist of a header section that defines all of the characteristics of the drawing environment and a body that contains the definition of the graphic objects. Figure 5.19 is a listing of part of the file that defines the graphical illustration in Figure 5.18. The x, y coordinates that outline the middle rectangle are listed below the ".5614 g" command that defines its shading. The value was entered in the program as a percentage black, but is converted to a percentage white in a decimal format in the saved document (i.e., 1.0 = white; 0.0 = black). The decimal value associated with the PostScript "g" command can have as many as four places. Thus gray shadings can be defined between .0001 and .9999, making the specification of 9,999 separate shadings between white and black possible. (The number of distinct shadings that are actually produced depends upon the resolution of the printer or imagesetter — see box on following page.)

The MapArt collection from Cartesia includes numerous polygon-based files in a PostScript format that can be used with Illustrator or Freehand. A portion of one of these files defining a single polygon is shown in Figure 5.20. The "Note" statement includes a combination of the FIPS codes of the state (31) and county (077) and the name of the county. The following six coordinates outline Greeley county (m indicates a move; l defines a line).

```
0.5614 g           ! =43.86 % ink; (1.0 -
.5614)
254 335.5 m        ! move to this point
254 353.5 l        ! draw to this point
223 353.5 l        !       "
223 335.5 l        !       "
254 335.5 l        !       "
```

Figure 5.19   Postscript commands that define a shaded rectangle. The first statement defines the shading. The second statement is a 'move' to the first point of the rectangle. The following four coordinates outline the rectangle.

How Many Shadings Can a Laser Printer Make?

Buttenfield (1993) gives the following formula for determining the number of gray shades that a laser printer can produce:

$$\text{Number of possible gray levels} = (\text{ Printer Resolution / Lines per Inch })^2 + 1$$

where printer resolution is the number of dots per inch (dpi) produced by the printer. Older laser printers print at 300 dpi whereas newer models generally print at 600 dpi or 800 dpi. Imagesetters that print on film normally have resolutions of 2,540 dpi.

The number of lines per inch influences the texture of the shading. The lower the line per inch value, the greater the texture in the shading (the individual dots are bigger). The higher lines per inch value will produce finer shadings but limits the variation in the size of the dots, producing fewer possible shadings (Leonard and Buttenfield 1989).

On a 300 dpi laser printer with a relatively course line per inch value of 42.5, a total of 50 distinct shadings are possible. At 70 lines per inch only 19 shadings are possible. However, by increasing the number of dots per inch of the printer, substantially more shadings can be produced. A 600 dot per inch laser printer can produce 200 shadings at 42.5 lines per inch and 74 shadings at 70 lines per inch. On a 2,540 dpi imagesetter up to 3,572 shadings can be created at 42.5 lines per inch and, 1,317 shadings are possible at 70 lines per inch.

Computing the shading value for a polygon is accomplished by rescaling the data on a 0 to 100 scale:

$$P = (z_i - zMin) * (100. / zRange);$$

where, P is the perceived value to be used in the perceptual adjustment formula, z_i is the data value, zMin is the minimum data value, and zRange is the difference between the maximum and minimum data values.

Assigning a shading value to a polygon is possible within Illustrator by selecting a polygon and specifying a shading value as a percentage ink. However, shadings can be assigned more quickly with a text processor. This is done by simply inserting the "g" command along with the corresponding shading value expressed as percentage white (as computed with the perceptual adjustment formula) following the "Note" statement. The "s" at the end of the polygon in the unshaded file must also be changed to a b (last line) to have the shading take effect. The listing in Figure 5.21 shows the addition of a shading value for Greeley County, Nebraska.

```
%%Note:31077,Greeley,NE  ! id of county
367.9533  385.0322  m         ! move to this point
403.29  385.631  l              ! draw to this point
403.29  358.0805  l             !      "
403.29  348.4977  l             !      "
368.5523  348.4977  l           !      "
367.9533  385.0322  l           !      "
```

Figure 5.20   The outline of Greeley County, Nebraska, defined in Postscript. The first statement is a note that includes the ID value and the name of the county. The coordinates that follow specify the outline of the county.

```
%AI3_Note:31077,Greeley,NE
0.6543 g                      ! the shading value
280.1552 53.6423 m
312.1679 54.1848 l
312.1679 29.2257 l
312.1679 20.5443 l
280.6977 20.5443 l
280.1552 53.6423 l
b                             ! end of shaded
```

Figure 5.21   Greeley County with shading value assigned. The 0.6543 g statement defines a shading of 34.57% ink $((1.0 - 0.6543) \times 100)$ for the polygon.

Figure 5.22 is an unclassed choropleth map of Nebraska showing the median housing value by county. The shadings were assigned using a text processor, and the file was saved in a text format. Illustrator was used to add the text, legend, and neat line and for printing the map.

5.6 SUMMARY

Computer programs for graphic design and graphic editing have had a major influence on cartography. They are being used in all types of map making. Most maps that are now published in newspapers, magazines, and books have been created with graphic design programs.

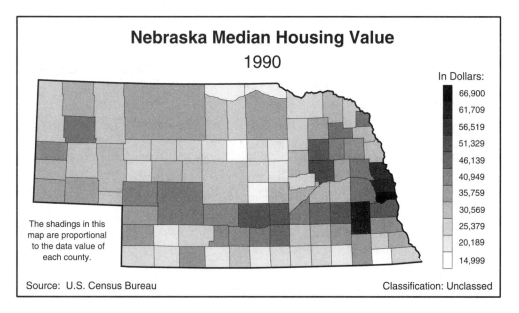

Figure 5.22   Unclassed choropleth map created with a text processor and Illustrator. The data values in this map have not been classified. The shading that has been assigned to each county is proportional to the data value. (Unclassed choropleth maps can also be created on the screen of a computer. Up to 256 gray shadings can be displayed with an 8-bit display.)

A variety of graphic design programs have been developed for the microcomputer. Some are based on a raster data structure, and others on vector data that is stored and accessed in the form of objects. The most widely used graphic processing programs in cartography are object illustration programs.

5.7 EXERCISES

1. Use a paint graphics program to create a map that shows where you live. The map should be designed for someone who is visiting you from another city. Include large scale as well as small scale depictions.

2. Create the same map as above but with an object program. What are the relative advantages and disadvantages of the two approaches for creating the map?

3. Create an unclassed choropleth map using a PostScript object program such as Illustrator. You may use the data from the fifth exercise in Chapter 4.

5.8 REFERENCES

BUTTENFIELD, B. P. (1993) "Formalizing Rules for Gray Tone Selection in Computer Mapping." *Proceedings of the American Congress on Surveying and Mapping* 1: 52–60.

DOBSON, M.W. (1973) "Choropleth maps without class intervals." *Geographical Analysis* 3: 358–360.

LEONARD, J. AND BUTTENFIELD, B. P. (1989) "An Equal Value Gray Scale for Laser Printer Mapping." *The American Cartographer* 16, no. 2: 97–107.

CARTESIA SOFTWARE. (1994) *MapArt.* Lambertville, NJ: Cartesia Software.

PETERSON, M. P. (1992) "Creating Unclassed Choropleth Maps with PostScript." *Cartographic Perspectives*, no. 12 (Spring): 4–6.

PETERSON, M. P. (1979) "An Evaluation of Unclassed Crossed-Line Choropleth Mapping." *The American Cartographer* 6, no. 1: 21–37.

TOBLER, W. (1973) "Choropleth Maps Without Class Intervals?" *Geographical Analysis* 3: 262–265.

WILLIAMSON, G. R. (1982) "The Equal Contrast Gray Scale." *The American Cartographer* 9: 131–139.

6

Image Processing

6.1 INTRODUCTION

Image processing, a form of graphic processing, involves the manipulation of digital images with the computer. Image processing is closely associated with remote sensing, where its use dates to the 1960s. However, it was not until the 1970s after the launch of a series of satellites called LANDSAT, that digital image data became widely available. The scanning of aerial photographs became another source of digital images. The display and processing of digital imagery, traditionally associated with workstations, is now being performed with microcomputers.

A major advantage of the computer is its ability to integrate the display of maps and images. The images may be from above, as with air photographs and satellite images, or they might consist of pictures taken on the ground. In both cases the images can help to create a better correspondence between a map and the reality it depicts. An interactive setting would make it possible for the user to switch between the map and the image. This could be implemented with a sliding bar, as depicted below, that would allow the two views to be faded in and out interactively.

Digital images also present possibilities for animation. The animation of satellite images is routinely used to display the movement of clouds as part weather forecasts on television. Other uses of animated satellite images include the creation of flyovers that simulate flying over a terrain.

111

There are two major approaches to remote sensing. The first, referred to as *image oriented,* is based on the human ability to recognize objects or spatial patterns. Air-photo interpretation is an image-oriented form of remote sensing that stresses object recognition. *Numerically oriented* remote sensing, on the other hand, emphasizes the inherently quantitative aspects of the digital image and uses the computer to enhance the imagery or identify spectral patterns. A spectral pattern results when an object reflects energy differently in the various bands of the spectrum. Satellites typically collect data in multiple bands of the electromagnetic spectrum.

Digital imagery can be a major component of the interactive and animated map. In this chapter we examine the characteristics of digital images, their display, the types of digital imagery, and methods of image manipulation and enhancement.

6.2 CHARACTERISTICS OF THE DIGITAL IMAGE

Much of the imagery that is collected for remote sensing is from parts of the spectrum that are not visible to the human eye. The entire range of wavelengths is referred to as the electromagnetic spectrum. Figure 6.1 represents parts of the electromagnetic spectrum that are normally used in remote sensing and the associated sensing device. The effective range of remote sensing is from 0.3 μm (microns; one micron equals one millionth of a meter) to 1 meter.

The human eye responds to wavelengths between 0.38 and 0.72 μm. Wavelengths below blue light, referred to as ultraviolet, are badly scattered by particles in the atmosphere and are therefore not useful in remote sensing. The part of the spectrum beyond red light, called the infrared, has proven very useful for a number of applications. For example, it has been found that healthy vegetation reflects a relatively high amount of the wavelengths in the near infrared part of the spectrum. Vegetation

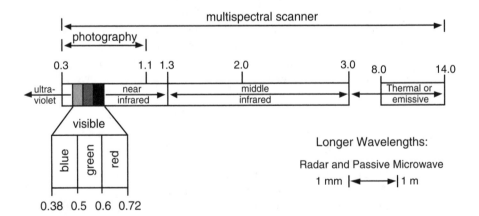

Figure 6.1 Portions of the electromagnetic spectrum that are used in remote sensing and the associated sensing device. Numbers indicate wavelengths of energy in microns (one millionth of a meter). The longer wavelengths associated with radar are between 1 millimeter and 1 meter.

that is under stress as a result of disease or lack of moisture reflects a smaller proportion of wavelengths in this part of the spectrum.

The near infrared and middle infrared parts of the spectrum, from 0.72 to 3.0 μm, are referred to collectively as reflected infrared. Atmospheric effects greatly complicate the usefulness of the wavelengths between 3.0 μm and 8.0 μm. Beyond this point, to about 14 μm, is the thermal infrared part of the spectrum, sometimes referred to as the far infrared. Energy in this part of the spectrum is emitted (heat) rather than reflected (light). The microwave part of the spectrum consists of very long wavelengths between 1mm to 1m. These wavelengths are not as prevalent in the environment making remote sensing in this part of the spectrum dependent on an active *radar* sensor that bounces energy off environmental features and records what is returned. Thermal and radar imagery can be captured both day and night whereas the visible and reflective infrared parts of the spectrum can be effectively imaged only in daylight.

Multispectral scanners capture a number of images simultaneously. Each image represents an energy response in a spectral band. A spectral band may consist of any portion of the spectrum. One could capture blue, green, and red light in the visual spectrum with three separate bands between 0.4 to 0.5, 0.5 to 0.6, and 0.6 to 0.7 microns. A single *panchromatic* band could also be used to image the entire visible spectrum. Often bands are chosen according to the reflectance characteristics of specific features. For example, a band between 0.63 and 0.69 microns corresponds to the chlorophyll absorption region for vegetation.

6.3 DIGITAL IMAGE DISPLAY

Digital images are stored in a raster format. Each individual pixel is usually represented with one byte. A byte consists of 8 bits, each representing the numbers 128, 64, 32, 16, 8, 4, 2, and 1. Figure 6.2 shows how 8 bits can represent all values between 0 and 255. The amount of reflected or emitted energy for an individual pixel is usually represented as a value between 0 and 255 (one byte per pixel). The higher the value, the greater the amount of reflectance or emittance.

The display of a single digital image can be done with gray shades. A pixel with a value of zero is depicted with black, 255 with white, and values between 1 and 254 with the appropriate shade of gray. To represent this many shadings, 8 bits of video memory are needed for each pixel on the screen.

The creation of a color image requires three separate digital images. Color is

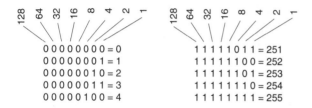

Figure 6.2   The representation of numbers with one byte (8 bits). A total of 256 distinct values from 0 to 255 can be represented.

represented on the monitor of the screen with the additive primaries of blue, green and red. A separate band is needed for each of these primaries. The first band would be represented with 256 shades of blue, the second with 256 shades of green, and the third with 256 shades of red, for a total of 16.7 million colors (2^{24}). To display an image on the screen, 24 bits or 3 bytes of video memory are required for each pixel. If the separate bands correspond to blue (0.4 – 0.5 μm), green (0.5 – 0.6 μm), and red light (0.6 – 0.7 μm), the resultant screen display represents the *natural colors* — the colors as we see them. If the bands do not correspond to these colors, we have the creation of a *false color* image. A false color image results when the blue, green, and red primaries are used to represent any other part of the spectrum.

A common false color composite assigns the band between 0.5 to 0.6 μm to blue, 0.6 to 0.7 μm to green and 0.7 to 1.1 μm to red. This particular assignment corresponds to a color infrared picture. In order to show the infrared part of the spectrum, the blue light (0.4 to 0.5 μm) is not depicted. A color shift takes place to incorporate the infrared reflectance into the color image (Figure 6.3). Because vegetation reflects more energy in the infrared band than in the red or green bands, a common characteristic of color infrared imagery is that vegetation appears red.

6.4 TYPES OF IMAGERY

Imagery of the earth can be classified by the medium — photography or scanner — and the platform — airplane or satellite. The platform for photography is normally the airplane, although satellites have been built for taking photographs, primarily for military applications. Satellites are the primary platform for scanners but it is also possible to scan images from an airplane. Photographic images can be more easily interpreted because they are usually associated with the visible portion of the spectrum. Scanner–derived images are often difficult to interpret, especially when they are combined to form color images.

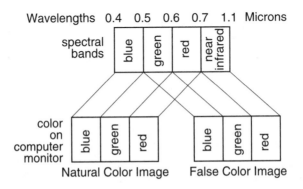

Figure 6.3   The distinction between a *natural* and *false* color image. The red color is used to depict the near infrared on a color image. A false color image results when the blue, green and red primaries are used to represent any other part of the spectrum. A color image that incorporates any band beyond visible light is by definition a false color image.

6.4.1 Aerial Photography

Aerial photography can be characterized by differences in film and photo scale. There are essentially four types of film used in aerial photography: panchromatic (black and white), panchromatic infrared, color, and color infrared. For reasons of cost, black and white is by far the most prevalent. Its main application is in the creation of stereo images that are used to derive elevation information for the creation of topographic maps. Color infrared is the more prevalent form of color photography in remote sensing.

Common photo scales in remote sensing range from 1:130,000 to 1:6,000. A number of federal agencies in the United States provide air photo coverage over large parts of the country (Table 6.1). The photographs are approximately 9 inches square (23 cm x 23 cm) and represent an area with approximately 30 km to 1.4 km on a side. The National Aerial Photography Program (NAPP), a multiagency program coordinated by the United States Geological Survey (USGS), provides coverage of the entire United States at a scale of 1:58,000 for color infrared and 1:80,000 for panchromatic. These two types of photographs are taken simultaneously at an altitude of 12,200 m.

All aerial photography is subject to a distortion caused by the central perspective of the camera lens. This distortion increases in a concentric pattern away from the center of the photograph. The periphery of an aerial photograph will normally have a smaller scale than its center. Another characteristic of this *radial distortion* is the outward displacement of the tops of tall objects from their bases. It is possible to remove this and other forms of distortion through the creation of an orthophotograph. Orthophotos, like maps, have one scale throughout. They are prepared with an orthophotoscope, which transfers all points of a stereomodel onto an orthophoto negative. The USGS also distributes digital quarterquad orthophotos in a raster format at a scale of 1:12,000.

Table 6.1   Types of aerial photography. Photographic imagery is available at a variety of scales. Some imagery is acquired on a yearly basis for large parts of the United States.

Photo Scale	Area per frame (KM)	Description
1 : 130,000	29.9 x 29.9	NASA high altitude photography
1 : 120,000	27.6 x 27.6	NASA high altitude photography
1 : 80,000	18.4 x 18.4	National Aerial Photography Program (NAPP), panchromatic
1 : 65,000	14.9 x 14.9	NASA high altitude photography
1 : 60,000	13.8 x 13.8	NASA high altitude photography
1 : 58,000	13.3 x 13.3	National Aerial Photography Program (NAPP), Color IR Film
1 : 40,000	9.2 x 9.2	USGS and USDA current mapping photography program
1 : 24,000	5.5 x 5.5	Photography that matches scale of USGS 7 1/2 'quads
1 : 20,000	4.6 x 4.6	Archival USGS and USDA mapping photography
1 : 15,840	3.6 x 3.6	USFS photography
1 : 6,000	1.4 x 1.4	EPA photography for analysis of hazardous waste

NASA = National Aeronautical and Space Admin.   USFS = United States Forest Service
USGS = United States Geological Survey   EPA = Environmental Protection Agency
USDA = United States Department of Agriculture   All photos approximately 230 x 230 mm.

6.4.2 Satellite Imagery

Satellite images can be differentiated on their spatial, spectral, and temporal resolutions. Spatial resolution refers to the ground distance that is represented by each pixel. Spectral resolution is the number and width of the spectral bands that are imaged. Temporal resolution is the frequency with which an image is made of a particular area.

The distinction between spatial and spectral resolution can be clarified by comparing the characteristics of the imagery from the US LANDSAT and the French SPOT satellites. There have been a total of five LANDSAT satellites since 1972. The first three satellites, all launched in the 1970s, included multispectral scanners that imaged four bands in the visible and near infrared parts of the spectrum with a spatial resolution of about 80 meters per pixel. LANDSATs 4 and 5, launched in 1982 and 1985, respectively (LANDSAT 6 was launched in 1993 but did not deploy successfully), includes an advanced multispectral scanner called the Thematic Mapper that images seven bands at 30 meters per pixel (Figure 6.4). SPOT, first launched in 1986, includes a

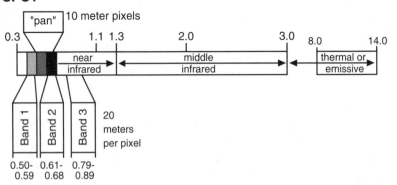

Figure 6.4   Spectral resolutions of the LANDSAT Thematic Mapper (TM) and SPOT multispectral scanners. TM has a higher spectral resolution because it divides the spectrum into more bands. The SPOT scanner has a higher spatial resolution at 20 meters per pixel compared to TM at 30 meters per pixel.

multispectral scanner with only three spectral bands but a spatial resolution of 20 meters per pixel. It also includes a panchromatic scanner (0.51 to 0.73 μm) with a spatial resolution of 10 meters per pixel (see Figure 6.5). In summary, LANDSAT TM scanner has a higher spectral resolution (because it divides the spectrum into more bands) while the SPOT system has a higher spatial resolution

Temporal resolution refers to the repeat cycle of a satellite. LANDSAT and SPOT are both polar-orbiting satellites that orbit the earth at less than 1000 km and image a relatively small swath of about 200 km on each pass. LANDSATs 4 and 5 image the same spot on the earth every 16 days. SPOT has a repeat cycle of 26 days. The *Advanced Very High Resolution Radiometer* (AVHRR) sensor is an example of a system with a courser spatial resolution but a higher temporal resolution (Figure 6.6). The scanner has been deployed on a series of polar–orbiting satellites from the National Oceanic and Atmospheric Administration (NOAA) and has a repeat period of between four and five days. Spatial resolution is much lower. Each pixel represents between 1 km to 4 km. AVHRR divides the spectrum into five bands (0.58 – 0.68 μm, 0.72 – 1.10 μm, 3.55 – 3.93 μm, 10.5 – 11.5 μm, and 11.5 – 12.5 μm). AVHRR imagery is used for the evaluation of snow cover, flood monitoring, vegetation mapping, soil moisture analysis, and fire detection (Lillesand and Kiefer 1987, 596).

Figure 6.5   A portion of a panchromatic SPOT image of Phoenix, Arizona. The spatial resolution is 10 meters per pixel. The airport is visible in the middle of the image and the central business district is toward the upper-left. The image was acquired on January 12, 1994.

Figure 6.6   AVHRR mosaic from a NOAA-12 satellite showing parts of North America, Greenland and the Atlantic Ocean. The three images that make up the mosaic were acquired in the morning hours of December 28, 1993. The time between each image is about 102 minutes. Notice the resultant discontinuity in the cloud patterns over the United States between the adjacent images. (Image from ftp site rainbow.physics.utoronto.ca.)

Images taken by geostationary weather satellites have the highest temporal resolutions. The *Geostationary Operational Environmental Satellite* (GOES) weather satellite, for example, generates an image every 30 minutes. Images from the GOES weather satellites, a part of almost any weather forecast on television, are the most commonly viewed satellite images. A network of geostationary meteorological satellites are positioned over the equator at a distance of 36,000 km (22,300 miles) and image the earth between 55°N and 55°S latitude. A GOES satellite positioned at 75°E longitude images the contiguous United States. Two parts of the spectrum from 0.55 to 0.90 μm and 10.5 to 12.6 μm are imaged. These two bands are called *visible* and *infrared* although the visible band includes a large part of the near infrared and the infrared band is actually in the thermal infrared part of the spectrum.

GOES satellite images provide information on both cloud thickness and height. The visible image depicts reflected sunlight. Thicker clouds reflect more of the sun's rays and therefore are distinguishable from thinner clouds. The infrared (thermal) band is used to determine the height of the clouds. The tops of higher clouds have a colder temperature than the tops of lower clouds. The colder clouds will appear light on an infrared GOES image, and the warmer clouds will appear darker (Ahrens 1991, 181). Storms are generally associated with higher clouds and would appear lighter.

A GOES image encompasses an entire hemisphere of the earth as seen from 36,000 km (Figure 6.7). These images undergo a considerable amount of processing at ground receiving stations. Subscenes of the image are extracted, and lines are added to represent the country and state outlines (see Figure 6.8). These processed images are then distributed via satellite. Animation loops are created by assembling a series of these images. In some cases the weather images undergo additional processing to display the heights of clouds in three dimensions.

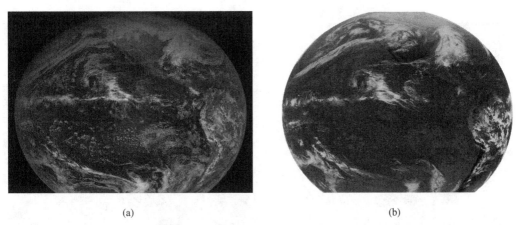

(a)   (b)

Figure 6.7   Full-disk GOES images from January 1, 1994 at 1900 Greenwich Mean Time. Image (a) represents reflected energy in a portion of the spectrum between 0.55 to 0.90 μm. Image (b) depicts emittance in the thermal wavelengths between 10.5 to 12.6 μm.

Figure 6.8   A portion of a processed GOES image of North America, August 2, 1993 at 20:00 Greenwich Mean Time.   The outlines of countries and U.S. states have been delineated.

Radar (*radio detection and ranging*) imagery is becoming a more common form of remote sensing. It is an active form of sensing that involves the transmitting of a pulse of energy and then recording the "reflections" or "echoes"of the return signal. There are a number of advantages to radar imagery. Radar images can be created at night. The microwave radiation that is used by radar is also capable of penetrating haze, light rain, snow, clouds, smoke, and dry soil (up to about 2 meters). Radar is especially useful in mapping terrain. However, the imagery is often difficult to interpret as one needs to be aware of the reflection-refraction characteristics of different surfaces. An example of a radar imaging device is Shuttle Imaging Radar (SIR), flown on three separate NASA shuttle missions and called SIR-A, SIR-B, and SIR-C (Figure 6.9).

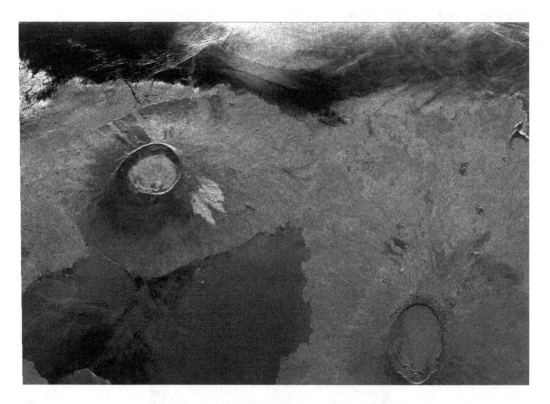

Figure 6.9   A Spaceborne Imaging Radar C (SIR-C) image showing part of Isla Isabella in the western Galapagos Islands.   The image is centered at about 0.5 degree south latitude and 91 degrees west longitude and covers an area of 75 by 60 kilometers (47 by 37 miles).   The radar incidence angle at the center of the image is about 20 degrees. The western Galapagos Islands, which lie about 1,200 kilometers (750 miles) west of Ecuador in the Pacific Ocean, have six active volcanoes similar to the volcanoes found in Hawaii. There have been over 60 recorded eruptions of these volcanoes. This image shows the rougher lava flows as bright features, while ash deposits and smooth pahoehoe lava flows appear dark. (From Internet site jplinfo.jpl.nasa.gov; see Appendix C.)

6.5 IMAGE MANIPULATION

Image manipulation encompasses a variety of numerical operations that are applied to a digital image. The techniques can be divided into four categories (after Lillesand and Kiefer 1987, 610):

1. Radiometric and geometric correction. Radiometric correction is the standardization of pixel values based on a reference value, and is usually done shortly after the image is received on the ground. Geometric correction removes spatial distortions arising from a number of sources. The sweep of the scanner, for example, leads to "tangential distortion" at the periphery of the image. In most cases these distortions are removed before the imagery is distributed.

A new class of software called *image morphing* is used to do the reverse of geometric correction, that is, it is a method for warping the shapes of object and images. Morphing is widely used in the film industry to dynamically change the appearance of a character (e.g., Terminator 2: Judgment Day). The software is normally used to intentionally alter the geometric characteristics of an image, but can also be applied to bring two images or an image and a map into congruence.

2. Image classification. Image classification is the basis of automated remote sensing and involves the identification of features based on spectral patterns. A spectral pattern describes the reflectance or emittance responses over a portion of the spectrum. Figure 6.10 depicts the spectral response patterns of natural grass and artificial turf over the 0.3 to 1.1 µm region of the electromagnetic spectrum. These two features can be discriminated as separate features because they have a much different spectral response in the near infrared part of the spectrum.

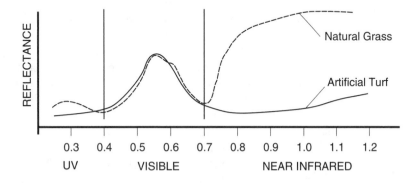

Figure 6.10   Spectral response patterns of natural grass and artificial turf. The two patterns are essentially identical in the visible portion of the spectrum, but in the near infrared portion, the natural grass reflects a higher amount of infrared. Image classification in remote sensing is accomplished by separating different spectral patterns.

3. Data merging. Data merging is the combination of different types of digital imagery or digital imagery and maps. An example of image merging is to combine a LANDSAT TM image with a SPOT, bringing together the differing spectral and spatial qualities of these satellite images. An image-map form of data merging would be the combination of a digital image and a digital elevation model (DEM). The process involves overlaying the image on top of a three-dimensional representation of the DEM.

4. Image Enhancement. Image enhancement is the manipulation of image contrast or the extraction of particular features such as lines. There are two major types of image enhancement, the first based on the manipulation of image contrast and the second on a spatial filtering operation that is used to accentuate features.

a. Contrast Enhancement

Digital images rarely encompass the entire range of possible digital values between 0 and 255 (8-bit encoding). The objective of contrast enhancement is to stretch the entire range of digital values or a portion of the digital values so that the resultant image has a greater contrast (Figure 6.11a). The formula for contrast enhancement is:

$$\text{New Pixel Value} = \left(\frac{\text{Original Pixel Value - MIN}}{\text{MAX - MIN}} \right) \cdot 255$$

where, MIN is the minimum pixel value in the original data and MAX is the maximum pixel value in the original data. It is also possible to stretch only a portion of the original image, as in Figure 6.11b. Here the MIN and MAX values would be the boundaries of the portion that is to be stretched, in this case 135 and 177.

A non-linear form of contrast enhancement is called histogram equalization. With this approach, the area with the higher frequency of pixel values is stretched over a larger part of the possible range (Fig. 6.11c). A relatively smaller part of the range is reserved for the less frequently occurring pixel values.

Threshholding and color slicing are two additional approaches that are used to manipulate the display of the pixel values. Threshholding converts the image into classes. For example, a threshold applied at the 127 level would convert values at 127 and above to white and below 127 to black. The resultant threshold could be used as an overlay to manipulate a particular portion of another image. Color slicing assigns a series of colors to a gray scale image. A specific range of gray values is assigned an arbitrary color. Multiple slices can be made to accentuate a number of gray-scale ranges. The procedure is sometimes referred to as colorization and can be done interactively in some programs by clicking along a histogram of the image.

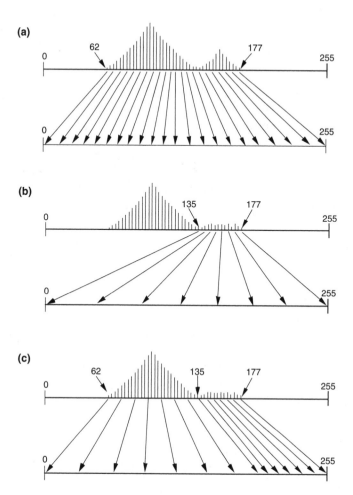

Figure 6.11   Examples of contrast stretching. In (a) the entire image is stretched in a linear fashion. In (b) only part of the original image is stretched. In (c), called histogram equalization, the high frequency part of the image is stretched in a non-linear fashion.

b. Spatial Filtering

The filtering of a digital image is a much different numerical operation than contrast enhancement. It is based on a procedure called convolution that manipulates an image through a 3 x 3 or larger kernel. This kernel or convolution matrix contains values that specify the adjustment to the middle cell. The calculation is performed repeatedly on a portion of the original image (Figure 6.13). Because of the way the calculation is done, the transformation cannot be applied to the edge pixels in the original image.

A relatively simple convolution is to have each cell in the 3 x 3 or larger matrix contribute equally to the new value corresponding to the middle cell. In the 3 x 3 case, for example, there are a total of nine cells so that each cell contributes equally to the value of the middle cell (Figure 6.12). The effect of this filter is to smooth the image (Hord 1982, 81).

Filtering is accomplished by altering the contribution of each of the surrounding pixels on the new value of the middle cell. For example, the horizontal filter accentuates vertical lines by multiplying the adjacent vertical pixels by the same negative and positive constants. The addition of these numbers then accentuates the difference in value between the adjacent vertical cells. A similar procedure is used with the vertical and diagonal filters. The Laplacian filter accentuates the difference between the center and surrounding cells. It has the effect of enhancing the edges in the image (an edge is defined as a relatively large difference in gray value between adjacent pixels; see Figure 6.13).

Many of the operations associated with image processing require considerable computing resources and are therefore still performed on UNIX-based workstations. But, smaller computers can now perform the same image manipulation operations. Images and the programs to manipulate them can be obtained through the Internet (see Appendix C). Examples of programs that are available through Internet include NIH Image, a product of the National Institutes of Health, Research Services Branch, and NCSA Image from the National Center for Supercomputer Applications. Both can read, display, edit, enhance, analyze, print, and animate images in a variety of raster formats.

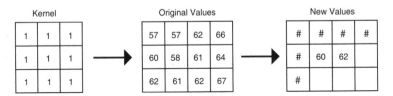

Figure 6.12   Spatial filtering of a digital image. The kernel specifies the type of filter that is applied. In the example on top the kernel specifies that each cell in the 3 x 3 portion of the matrix contributes equally to the value of the new middle cell. Kernels that constitute other filter types are included at the bottom of the figure.

Figure 6.13   An example of spatial filtering. The original image is on the left. The image on the right has been filtered with a Laplacian 5 x 5 filter and contrast stretched with histogram equalization. The filter accentuates small differences in gray-scale values. The filtered image shows variations in the gray values that are not evident in the original image, especially in the ocean areas.

They support many standard image manipulation functions, including both contrast enhancement and spatial filtering.

6.6 SUMMARY

Imagery of the earth can be acquired through photography, multispectral scanners, or with radar. Photography is sensitive to the visible and near infrared parts of the electromagnetic spectrum. Multispectral scanners can image the visible, near infrared, middle infrared, and thermal parts of the spectrum. Radar is an active type of remote sensing that can be used at night and is immune to many of the atmospheric effects that plague other forms of remote sensing. Regardless of the source, all imagery can be represented in digital form.

A digital image can be manipulated in a number of ways. Contrast stretching is a manipulation of the gray values to improve image contrast. Image filtering, based on a kernel, can be used to accentuate different aspects of the image. Other image processing operations include geometric correction, morphing, and image classification.

Images of the earth, either from a satellite or an aerial photograph, provides a valuable view of the earth. For most this view is unfamiliar, however, combined with maps in an interactive setting, this type of imagery can take on greater meaning and become a valuable addition to the presentation of maps.

6.7 EXERCISES

1. Image editing programs (e.g., Photoshop) provide tools to manipulate individual pixels and enhancement tools to manipulate the image as a whole. How many types of the contrast enhancement procedures can be performed with these programs?

2. A number of programs can be obtained through the Internet for image processing (e.g., NIH Image and NCSA Image). Compare the functions of these programs.

3. Raster-based GIS programs (GRASS, IDRISI, MAP II) incorporate a number of image manipulation functions. What type of imagery can these programs handle, and how can images be manipulated?

6.8 REFERENCES

AHRENS, D. C. (1992) *Meteorology Today: An Introduction to Weather, Climate, and the Environment.* St. Paul, MN: West Publishing.

LILLESAND, T. M. AND KIEFER, R. W. (1987) *Remote Sensing and Image Interpretation.* New York: John Wiley.

HORD, R. M. (1982) *Digital Image Processing of Remotely Sesnsed Data.* New York: Academic Press.

7

Multimedia and Hypermedia

7.1 INTRODUCTION

Multimedia is the various combinations of text, graphics, animation, sound, and video for the purposes of improving communication. At one time a multimedia presentation was simply a slide show and an accompanying narration on tape. The use of the computer has added the component of interactivity, in which the user is actively engaged in the selection and presentation of information, not simply a passive observer of a fixed procession of sight and sound. This form of multimedia is sometimes referred to as interactive multimedia.

Hypermedia is a particular form of interactive multimedia. *Hyper* refers to non-linear methods of moving through a body of information, as opposed to a paper document or a slide-show multimedia presentation that forces the observer to move in a structured direction. One of the first and most articulate proponents of what is now called hypermedia was Vannevar Bush, science advisor to President Franklin D. Roosevelt in the 1940s. In a 1945 article entitled "As We May Think," Bush foresaw a time when data communications, automation, and miniaturization would combine to aid thinkers in every area of study. He envisioned a Memex device designed around the concept of association.

> Consider a future device ... which is sort of a mechanized private file library. It needs a name, and, to coin one at random, "memex" will do. A memex is a device in which an individual stores his books, records, and communications ....

It is an enlarged intimate supplement to his memory.
... The human mind ... operates by association. With one item
in its grasp, it snaps instantly to the next that is suggested by
association of thoughts, in accordance with some intricate
web of trails carried by cells of the brain ...

Man cannot hope fully to duplicate this mental process
artificially, but he certainly ought to be able to learn from it ...
Selection by association, rather than by indexing, may yet be
mechanized. (Bush, 1945)

Because of its emphasis on association, hypermedia can be viewed as a mechanized
support tool for thinking. According to Bush, users would be able to record and annotate
items and link them by "trails" that could be saved. He even speculated on the
possibility of voice input. With the proliferation of more powerful personal computers,
Bush's ideas are being realized.

Hypermedia had its formal beginning as hypertext in the early 1960s (the use of
other media on the computer was not yet feasible). Hypertext systems attempted to bring
Bush's concept of association into a text document. The basic unit of information in a
hypertext document is called a *node*. A node is a part of the document that covers one
concept. It may be an entire screen, as small as a word or as large as an entire book.
Nodes are connected to each other by electronic cross references called *links*. The
existence of a link is indicated by a *link anchor,* which may be a button or a boxed-in
word. When the "reader" activates the link anchor with the mouse, the screen
immediately displays the contents of that node. The user can branch off in different
directions, as desired. A hypertext document has no beginning or end. Computer-based
help systems generally implement a hypertext structure.

The nonlinear nature of hypertext presents the problem of becoming lost in the
maze of links (Horn 1989). This is called being "lost in hyperspace." The sequential
quality of ordinary text provides a structure for ideas with one idea following another.
Hypertext authors have been able to incorporate some visible structure in their works by
providing a map that depicts the links taken to a certain node. The map provides
information so that people can "find themselves" in hyperspace. Much of the work in
hypertext has been experimental. A program that implements most of the elements of
hypertext is Hyperties (Cognetics).

The mind appears to work by association. One idea or image gives rise to other
ideas and images. Hypermedia attempts to take advantage of this ability to associate
material. The trend toward interaction in multimedia has led naturally to the
incorporation of a more sophisticated linking structure. Hypermedia concepts are being
incorporated within computer-based multimedia, with some now equating hypermedia
and interactive multimedia. The hypermedia concept reminds us that it is the linking that
is important, not simply the integration of multiple media.

With the growing interest in interactive forms of communication, multimedia and
hypermedia have become major areas of research and development. The objective of

this effort is to revolutionize how information is communicated. The goal is to transcend the static and sequential character of the printed page. Computer programs for multimedia are beginning to be used in cartography. Electronic atlases are an attempt to add multimedia and hypermedia techniques to the display of maps. One function of maps is to create a *conception of place*, a feeling of being there, something maps on paper rarely accomplish. Multimedia and hypermedia have the potential to create this conception through the integration of multiple media and sophisticated linking structures.

The combination of maps and other media will have a major influence on maps and their use. In this chapter we examine the different types of media, their processing and storage on the computer, multimedia authoring, and the supporting medium of CD–ROM.

7.2 THE ELECTRONIC MEDIA

Media are the basic form of storing and communicating information. The five basic types of media that can be incorporated within multimedia are text, drawings, pictures, video, and sound. The unique combination of these media is at the heart of multimedia and hypermedia.

7.2.1 Text

The display of text on the computer would seem to be a simple task. However, the number of different lettering fonts, sizes, and styles provide a wide choice in text display options. Adding motion, like the rolling credits in a movie, provides an even greater array of display possibilities. In addition, programs are available for the creation of 3-D effects with text (e.g. Pixar's Typestry and Strata's StrataType). These programs "extrude" text into the third dimension and then apply a rendering process that adds texture to the surfaces of the extruded letters. The overall objective of text display is legibility. Often the different text display options can hinder the reading of the text.

The objective in hypermedia is to implement links for different text elements. This is usually accomplished with a button that is located behind a word or a phrase. Clicking on the button branches off to another node. In general, the use of text in graphic-based multimedia displays is limited to short text segments. A graphic illustration can often be more effective in communicating a concept than text.

7.2.2 Drawings

Drawings, essentially composed of points, lines, and areas, can be created with paint and object graphics programs, as described in Chapter 5. There is at yet no universally accepted graphics file interchange standard (Kay and Levine 1992). On Macintosh computers, PICT and PICT2 (color) are standards for exchanging graphic illustrations.

These files can contain both paint and object graphics. Many of the "standard" file types are associated with output to a laser printer, such as PostScript and encapsulated Postscript (EPS). Programs for multimedia are able to import most of these file types.

7.2.3 Pictures

Pictures are stored in a raster format. Each line is composed of a series of pixels, or picture elements. An individual pixel is encoded with a certain number of bits that represent either varying degrees of gray or color. With 8 bits, or 1 byte, up to 256 different shades of gray or color can be represented. A full-color picture, with 1 byte or 256 shades of each of the three additive primary colors of red, green, and blue (RGB), requires 24 bits per pixel. It is also possible to have 36-bit images with 12 bits or 4,096 shades for each of the three primary colors.

A number of formats have developed for the storage of pictures. Probably the most common format is called GIF (Graphics Interface Format, pronounced "jiff"), developed by CompuServe (an electronic information service) to be accessible by a variety of computers. Accessibility is achieved through "reader" programs that are available for the different computers (e.g., Giffer). These "shareware" programs display a GIF file on the screen of the computer.

The Tagged Image File Format (TIFF) was developed to help link scanned images with programs for desktop publishing. A scanner is used for the input of graphics in a raster format. Slide scanners are commonly used for the input of pictures. The current TIFF standard supports three main types of image data: black and white, gray scale and color (8-bit and 24-bit). It uses several compression methods with ratios ranging from 1.5:1 to 20:1, depending upon the amount of white space.

PCX is a raster graphics file format that handles monochrome as well as 2-bit, 4-bit, 8-bit and 24-bit color. It uses a run-length encoding scheme (RLE) to achieve compression ratios of 1.1:1 to 1.5:1 (see Chapter 4). Images with large blocks of solid colors compress best under the RLE method.

Another format is based on a picture file compression standard that was developed by the Joint Photographic Experts Group (JPEG). The JPEG standard can achieve compressions of 25 to 1 with only a slight loss in picture quality. JPEG is usually used for color pictures because of the increased amount of data in these files. A number of shareware programs are available to view JPEG files (e.g, JPEGView).

TGA is a raster graphics file format used by TrueVision's Targa video graphics hardware. Targa is a PC-based video graphics board for high-resolution imaging. The board specializes in the manipulation and display of 16-, 24- and 36-bit color images.

Encapsulated Postscript, or EPS, is a graphics file format that contains both a Postscript encoding for printing and a TIFF or PICT image for preview on the screen. The Postscript code drives the printer, and the preview formats allow the image to be displayed and manipulated. Graphic formats that incorporate multiple resolutions are becoming increasingly common.

The Kodak PhotoCD is bridging the gap between photography and the computer. Pictures can be placed onto a CD disc when they are developed or from existing negatives or slides. A PhotoCD disc holds over 100 images at different levels of resolution (res levels). The images can be viewed on a television screen with a special PhotoCD player, or they can be read into the computer through a PhotoCD compatible CD-ROM drive. The disc contains five scans of each image: 128 x 192 pixels, 256 x 384 pixels, 512 x 768 pixels, 1024 x 1536 pixels, and 2048 x 3072 pixels. All five of the images are contained in one file, called an image pack. Each image pack is compressed with Kodak's proprietary *lossless* (minimum visual data loss) compression scheme. A typical image pack for a single picture consumes approximately 4.5 MB on the CD disc.

7.2.4 Video

Of the five basic types of media discussed here, video is the most powerful and persuasive form. Conventional video, normally stored on tape, is accessed sequentially and is only interactive through the limited start, stop, forward, and rewind options of the tape player. Bringing interaction to the display of video sequences has been a major advance in computer technology.

There are basically two approaches for displaying video with the computer: (1) interactive videodisc and, (2) digital video. The two differ in how the image is stored. The videodisc uses an analog form, as on videotape, whereas the digital video stores the image as a digital representation.

Interactive Videodisc

Available since the late 1970s, interactive videodiscs were slow to gain acceptance because a separate videodisc player was required. With the fall in the price of videodisc players and videodisc mastering, videodiscs have become one of the most widely used technologies in multimedia, with an extensive collection of videodisc titles.

Physically, a videodisc is a larger version of the audio CD. It uses a similar optical storage technology in which microscopic "pits" are embedded in the surface, but encodes the video and audio in an analog form. There are two basic formats. Movie videodiscs that are sold to the home video market are in CLV (constant linear velocity) format and can hold one hour of video per side. The second format, CAV (constant angular velocity), is commonly used in interactive video applications. It can hold 54,000 still pictures per side along with up to a half hour of CD-quality stereo sound. On a CAV videodisc each frame of video is given a unique identifying number (1 to 54,000). These numbers are used by the computer to tell the videodisc player which frame number to display as a still image, or where to start and stop a video sequence.

The major advantage of videodisc over videotape is the ability to randomly access both still frames and video sequences. The amount of time required to search for a particular frame or sequence is usually less than three seconds. Picture quality is also good compared to the often jittery still images produced by videotape players.

A typical interactive videodisc workstation consists of a microcomputer, a videodisc player, and a video monitor or television set. The computer and the videodisc usually communicate through the "modem" port of the computer. Because the computer's function is simply to control the videodisc player, the computer does not need a color monitor or a very fast processor. The interactions between the computer and videodisc are relatively simple, centering around the indexing and integration of the large number of images.

Digital Video

Digital video became possible on microcomputers in the early 1990s. The technology makes it possible to display video clips directly from the hard disk of the computer with no additional hardware.

Digital video is accomplished through video digitizing and compression techniques that convert color video to a file that can be stored on disk. The tremendous amount of data used by a video sequence makes the compression of data necessary. A screen-sized frame of 24 bits of color requires about 900 KB of storage (640 x 480 x 3 bytes per pixel), and there are thirty such frames in every second of normal video. One second of video would require the processing of 27 MB of data, and this does not include the audio track. Compression schemes remove the superfluous or repetitive elements of the original video to create a more "economical" version. For example, two adjacent frames of video may be exactly the same and there is, therefore, no need to store the frame twice. This is essentially the technique used with the soon-to-be introduced high-definition (digital) television (HDTV). Other compression techniques reduce the amount of data required to describe each frame. A common compression technique called JPEG (Joint Photographic Experts Group) can achieve image compression rates of up to 25-to-1. Rates of 5-to-1 are more common because the images need to be quickly reconstructed when played back. Higher compression rates tend to slow down the playback because more time is required to reconstruct the individual images.

There are four major digital video schemes for microcomputers: Apple's QuickTime, Intel's Digital Video Interactive (DVI), Microsoft's Video for Windows, and MPEG (Motion Picture Experts Group). Both QuickTime and Video for Windows are extensions to the operating system that allow playback of a digital video file from within programs. Intel's DVI is a hardware solution to digital video. It includes a set of processor chips that compress video onto disk and decompress it in real time for playback at the U.S. standard video rate of 30 frames per second. Microsoft's product supports the AVI (Audio Video Interleaved) format and three video compression methods. AVI combines standard waveform audio and digital video to produce animations at 15 fps with a resolution of 160 x 120 x 8, and requires specialized hardware to work adequately. MPEG is an image compression scheme for full-motion video proposed by the Motion Picture Experts Group working in conjunction with the International Standards Organization (ISO). MPEG, like other compression schemes, takes advantage of the fact that full motion video is made up many successive frames consisting of large areas that are not changed. This digital video format stores up to 1 hour, 14 minutes of full-screen, full-motion video and synchronized CD-quality audio onto a CD-ROM disc. QuickTime has been more widely used because it is licensed in various forms to developers working in other operating environments, including Microsoft Windows and UNIX systems.

QuickTime

QuickTime movies appear as a separate window on the computer screen. They can be up to 640 x 480 pixels and achieve a video display rate of 15 frames per second. At 320 x 240 pixels, the frame rate increases to 30 per second. Additional hardware makes the display of MPEG digital movies within QuickTime possible. QuickTime also incorporates timecodes that enable it to synchronize sound with video. The sound portion may also be stored using MIDI code, an instruction set for the playing of music.

The maximum frame rates for playback of QuickTime movies varies with the model of computer and the compression method. One of the functions of QuickTime is to resolve differences in the playback speed of different computers. When playing audio or video clips, the software acts as a time juggler, matching the speed of the hardware to the demands of the audiovisual data. For example, if a video clip is recorded at 30 frames per second but the computer can only run the movie at 15 frames per second, QuickTime drops every other frame to keep the pictures and the soundtrack properly synchronized. (For further information on QuickTime, see Apple Computer 1993; Miller and Harris 1993; Drucker and Murie 1992; Hone 1993).

A QuickTime movie has the advantage of being able to be played from within any application, and be cut and pasted like text or graphics. A videoclip can be contained within a text document and played as a movie or examined frame-by-frame. The individual frames need not be from a video source but can be created in an illustration or image processing program and placed into a QuickTime movie with specialized software.

Digital video is not currently an alternative to the videodisc for multimedia applications that require long segments of high quality, full-screen video or large data bases of still images. For such video-intensive tasks, videodiscs are more cost-effective, even though additional hardware is required. Digital video can be used primarily to add short clips of video or graphic animations to presentations. Even with the current limitations of image size and playback, digital video is a dynamic new tool for audiovisual computing.

Video Input

Input from a video source is accomplished with a video digitizer, a hardware device that connects to the computer and to a VCR or camcorder. Video digitizers can be obtained for a relatively low cost and come standard with some computer systems. They may also incorporate hardware video compression to convert the video to a more compact form.

There are various forms of video. NTSC (National Television Systems Committee) is the format that is used in the United States and Japan. Most European countries use a system with a slightly higher resolution called PAL. A third system, SECAM, is used in France and a few other countries. Most video digitizers will accept both NTSC and PAL input.

NTSC and PAL have a so-called S-video version that has a better resolution than the standard form. An ordinary NTSC frame is composed of about 250 lines. NTSC S-video has about 400 lines, and even more lines are available in PAL S-video. Most televisions and video tape players still have the lower resolution, although some of the newer 8 mm videocameras are available that record in a "Hi" or S-video format. Slightly more expensive video digitizers are available that will accept S-video input.

Another source of video is the still video camera that records individual pictures on a small floppy disk. These come in low- and high-resolution versions, with 25, low-resolution pictures (260,000 pixels) and 50 high-resolution pictures (410,000 pixels) fitting onto a single disk. The cameras also output a video signal that can be digitized with a video digitizer.

Most video-based multimedia presentations are created with a video editor that does not require a conversion to digital video. The computer is an integral part of this process but is used only to keep track of the video sequences for editing purposes. The SMPTE and VISCA protocol are two alternative methods of controlling such editing functions, both using time codes on video tape. The various start-stop time codes can be made part of a data base to automatically control the editing of a tape. The same type of protocol can be used to designate video sequences for digitizing.

7.2.5 Audio

The use of sound in multimedia can either be as a background (like elevator music) or for the communication of information. Many of the sounds made by the computer are specifically designed to improve the user interface. The clicks made by the keyboard or the beep that accompanies an incorrect command are sounds that have informational content. The concept of "earcons," analogous to icons, has been proposed as one possible use of sound (Blattner, et al. 1989). Sound is just beginning to be explored as a means of communicating information with computers and maps (Blattner, et al. 1992). Whether background or "foreground," sound is an important part of multimedia.

There are four options for incorporating sound into an interactive multimedia application. The four approaches differ in how the sound is represented and stored.

Audio CD

The compact disc or CD has become the preferred method of publishing music, replacing the once-popular phonograph record. The two media differ in how sound is represented — the phonograph being analog, whereas the CD is a digital encoding. The phonograph works through the analog representation of sound as a "carved sound wave" within the record groove. On the CD sound is converted into digital code by sampling the sound waves 44,056 times per second and converting each sample into a 16-bit number. About 1.5 million bits of storage are required for each second of stereo sound, with every 16 bits encoding one of over 65,000 possible sounds (2^{16} or 65,536). Instead of a needle vibrating in the carved groove, a CD player shines a laser onto the microscopic pits, and resulting reflections are decoded to produce the sound.

Each musical piece on a CD is designated by a track number. CD players allow the user to choose a specific track to play. More sophisticated players make the selection of parts of a track or pieces of a song possible. The computer is used to control which parts of the CD will be played. The major limitation to the use of CD sound in multimedia is the inability to edit the sound. For example, it is not possible to change the speed of playback or to alter any parts of the soundtrack. One is also limited by the recordings that are available.

MIDI

MIDI stands for Musical Instrument Digital Interface, and like the CD it requires the use of a separate hardware device. The similarity ends there. MIDI is in essence a score containing instructions on how to play a piece of music. These instructions are passed through a MIDI interface box to a MIDI-compatible synthesizer, such as an electronic piano. The synthesizer's sound is then output to speakers by way of a separate audio amplifier. MIDI has become a standard for providing a common language that allows computers to communicate with all MIDI devices, including synthesizers, drum machines, and traditional instruments such as electronic pianos.

MIDI also provides recording capabilities. While an instrument is being played, the instructions can be recorded by the computer. This "recording" includes information about which keys were hit, how hard and fast the musical passage was played and other descriptive information. With MIDI, music that is created with any program, on any computer will play the same note, pitches, and rhythm when transmitted to any type of synthesizer.

Two hardware devices are required to play and record MIDI music. The MIDI interface box converts data from the computer (through a serial port) into a format that can be played by a MIDI device. These devices vary in the number of input and output port. The more expensive of these interface boxes can merge, synchronize, and process MIDI signals. The other required hardware device is the tone generator or synthesizer that actually produces the sounds. Inexpensive tone generators, about the size of a car stereo and weighing two pounds, can synthesize 192 acoustic and electronic instruments plus special effects and drum sounds, and can play them individually or in groups of 16 at a time.

MIDI software includes functions analogous to a cassette tape recorder — play, stop, fast forward, rewind, and record. A sequencing function allows sound to be edited and layered in multiple tracks. Each track can be independently transposed, shortened, deleted, or otherwise modified. The files produced by MIDI software are relatively small, with one minute of standard music encoded to about 45 KB.

Digital Audio

While MIDI is a score with instructions on how to play back sound, digital audio is a recording of the sound itself. Through a process called "sampling," sound is recorded and played back with every nuance of the original. The size of the digital audio file can be enormous, even after compression.

In digital audio, sound is sampled at specific intervals. Although sound is a continuous phenomenon, it must be recorded in discrete segments by the computer. The standard sampling rates are 44, 22, and 11 kilohertz (kHz), kilohertz being the frequency in thousands of cycles per second that a sound is captured. An adequate sampling rate to capture sound is twice its frequency. Our highest hearing frequency is about 17 kHz which would require a sampling rate of at least 34 kHz. Compact discs (CDs) are sampled at 44 kHz. Lower sampling rates will leave out sounds that we normally hear.

There is a trade-off between quality of sound and the amount of memory required to store the sound. At 22 kHz — the standard digital audio resolution — there will be 22,000 samples taken every second. If the sound is stored in 1-byte (8 bits) chunks of computer memory, then 22,000 bytes, or about 22 KB, will be needed for every second. Ten seconds will consume about 220 KB, and one minute requires about 1.3 MB. Twice this much space will be needed if the sound is stored at 16-bit (2 bytes) CD quality. Double this amount again for a stereo recording in which two separate tracks are recorded. A stereo, 16-bit CD-quality digital audio, sampled at 44.6 mHz, needs 10 MB per minute.

The difference between MIDI and digital audio is analogous to the difference between PostScript and bit-mapped image files. The PostScript file requires less space than the bit-mapped version because mathematical descriptions are used to describe the image. MIDI's formula approach is more economical that that of digital audio, where thousands of samples are captured each second. The contrast is quite dramatic: one minute of stereo sound with MIDI requires 45 KB (0.045 MB) versus 10 MB for digital audio. The savings ratio for MIDI is about 222 to 1.

The advantage of digital audio is that no other equipment is required for recording, playing, or editing the sound. On a computer with built-in sound, 8-bit digital audio can be recorded and played back using the internal microphone and speakers. For many applications short 8-bit sound segments are sufficient. Programs create a visual display of the waveforms of the sound (Figure 7.1), and any part of the sound can then be edited or played by interactively selecting a portion of the wave. In this case the sentence "How are you?" was recorded at a sampling rate of 22 kHz and stored at 1 byte per sample. This recording consumed about 80 KB.

Internal Sound Synthesizers

Many computers have a built-in sound generator or synthesizer that can be programmed to play music in a manner similar to the MIDI approach. The Macintosh includes a specialized sound chip that is capable of simulating multivoice music in stereo.

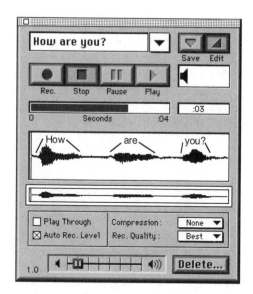

Figure 7.1   Audio dialog in HyperCard with portion of sound highlighted. The sound can be edited by selecting a part of the sound spectrum.

Sound generators are capable of synthesizing human speech. The basic building blocks of speech, called phonemes, are approximated by these chips. Words and sentences are put together by stringing together a series of phonemes. The speech produced in this manner can sound robotlike and may be difficult to understand.

The computer requires the user to specify the phonemes that make up speech sounds. To do this from scratch one must first learn the basic phonemes. There are approximately 60 for the English language. For example, the phonetic spelling of *heavy* is *h, eh1, v, y*. Software can be used to translate normal text to the basic phonemes. These programs cannot deal with the inflections and differential emphases that are an important aspect of speech. However, this form of synthesized speech requires far less memory than digitized speech. Like digitized speech, it can be incorporated as a sound channel to supplement or explain what the viewer is seeing on the screen.

The internal sound capabilities of computers can be used for many multimedia applications. The sound chips are improving but will never approach the sound capabilities of external MIDI devices for music. It is likely, however, that MIDI devices and more sophisticated speech chips will soon be incorporated within computers.

7.3 MULTIMEDIA AUTHORING

Multimedia authoring is the integration of media for the purpose of communication. A multimedia work revolves around a theme, similar to a story, and is authored in much the same way. If the work is organized sequentially, there is usually an introduction, a body, and a conclusion. Sequential multimedia presentations are the easiest to create because they simply move from one display to the next.

In nonsequential multimedia the user is able to make decisions about the course of events. This means that the multimedia work must incorporate a branching structure and a mechanism to respond to user choices. This mechanism is called scripting. Scripting is a form of programming but is easier to learn than standard procedural languages like FORTRAN, Pascal, or C. Scripting creates an interactive environment in which the user selects the order in which to view the information. Most multimedia authoring programs can be categorized into three types: presentation programs, card- or page-based authoring, and media integration.

7.3.1 Presentation Programs

Presentation programs are used mainly for the sequential display of text and graphics. The programs essentially emulate a slide show with special effects or transitions added between the individual frames. They include an outline editor that helps in the creation of text-based frames. The graphic material is often imported from other programs, although a graphic editor is usually a part of the program. Pictures and sound may also be incorporated, and the programs can display digital movies. PowerPoint (Microsoft), Persuasion (Adobe), and Freelance Graphics (Lotus) are examples of programs in this category.

Presentation programs are widely used in education and business. The simple editing of a file to alter a presentation is a major advantage of these programs. It is also possible to make a print-out of the presentation for distribution to an audience. This print-out contains snapshots of each frame and can include a place to add notes. In addition, presentation programs can be used to present a limited type of animation by displaying a series of frames in quick succession (called frame-based animation; see Chapter 8). The sequential character of these programs generally requires that the presentation be viewed from beginning to end, although it is possible to skip a specified number of frames.

7.3.2 Card-based or Page-based Authoring

Card-based or Page-based multimedia authoring programs are organized around a frame structure. Each "card" or "page" holds a specific amount of information that may consist of text, graphics, audio, or video. The most important characteristic of these programs is the ability to create links between cards and sometimes between elements of a card. A link is usually established through buttons that respond to user input with the mouse. Clicking on a button changes the display to another card as specified by a script associated with the button. A combination of several cards is stored as a file and would be organized around a particular theme. Examples of programs in this category are HyperCard (Claris), SuperCard (Allegiant), and Toolbook (Asymetrix).

These programs are designed for user interaction. The linking concept encourages multiple branching options. The user is given a choice of moving forward or backward within an area of interest or to a completely different topic. Animations can be shown by flipping through a series of cards. These programs are also commonly used for presentations but are designed for interactive use and the nonsequential display of information.

7.3.3 Authoring with Media Integration Programs

Media integration programs are the most sophisticated of multimedia programs, accepting a wide variety of different graphic and animation file formats. They incorporate a graphics editor and allow the direct input of video and sound (analog or digital). The programs incorporate a scripting language and can work with external programs. Finally, these programs can create stand-alone applications. Programs in this category include MacroMedia Director (MacroMedia) and Authorware (MacroMedia).

Media integration programs work by combining separate tracks of media. This integration occurs through a graphical user interface consisting of a grid with tracks layed out along the horizontal dimension and time along the vertical. Material is placed within each track and can be overlayed with other material in an adjacent track. Separate tracks exist for text, graphics, video, and sound. These programs can be used for presentations and for the creation of animations but are also designed for interactive use.

7.4 THE MEDIUM OF MULTIMEDIA — CD-ROM

Multimedia applications produce large files, especially those incorporating pictures, video, or sound. In order to distribute multimedia works a medium is needed with a large data storage capacity and low duplication cost. With a storage capacity of over 600 MB and a duplicating cost of under $1 per disc, the CD-ROM has become *the* medium of multimedia (Brown 1986).

The CD-ROM disc is the same size as the CD audio disc but uses a different track format. Unlike the standard computer hard disk, the CD-ROM is read-only; data can be written to it only once. CD-ROM drives are becoming a standard peripheral with microcomputers; the newer models have a faster search and read mechanisms and the ability to find and transfer data much more quickly than the original drives. Video segments can be transferred at reasonable viewing speeds. However, most video sequences available on CD-ROM discs are designed for viewing with the older, slower CD-ROM drives. Thousands of multimedia titles are available, ranging from encyclopedias to atlases. As with any new medium, the quality of these works varies considerably.

The way that data is stored on most CD-ROMs is based on the International Standards Organization (ISO) 9660 standard. Discs specifically designed for Macintosh computers use the Apple Hierarchical File Structure (HFS) format, but the Macintosh can also read the 9660 standard. Other specialized CD formats and file types have been developed for the Sony Data Discman and Sony CD-ROM/XA player. These are portable CD-ROM readers that are used mostly for text applications because of their limited screen resolution. The most common specialized CD format is the Kodak PhotoCD that is used for the storage of photographs.

Retrieval programs are usually included on the CD-ROM disc but may only work with one type of computer. These programs process the user's request, consult an index file, locate the requested information, and display it on the screen. The programs also compensate for the CD's slow access by minimizing the number of seek operations, and are designed to retrieve data as rapidly as possible. The traditional navigation aids available in printed works are also incorporated in CD-ROM retrieval software: table of contents, alphabetical indexes, page numbers, chapter, section, and subsection hierarchies. Some retrieval software allows the user to attach electronic notes in the margins.

CD-ROM publishing is comparable to the publishing of a book. The material is first assembled on a computer with a hard disk (capacity greater than 700 MB). Authoring software is used to create a user interface in order to access the material with a retrieval program (usually bundled with the authoring software). In a simulation step the work is tested for ease of use, and the layout of the material on the disc is optimized for fast retrieval. The prepared data files are written to 8mm "Exabyte," or 4mm "DAT" computer tape, removable optical disk cartridges or magnetic disk cartridges, and submitted to a replication facility. The first step in the multistep mastering process involves the creation of a glass master disk. The master disk produces the stampers that actually press the digital pattern onto the disc. Replication, also called pressing or stamping, is done at specialized compact disc plants that usually also reproduce audio CDs.

The mastering and replication process costs about $2,000 for a limited number of discs. However, acquiring content, conversion of data formats, editing, preparation for the CD-ROM, and licensing the retrieval software for inclusion on each disc will usually be far more expensive than the mastering and replication steps. The inclusion of multimedia elements, such as audio, video or computer animation, can introduce a whole new list of costs.

7.5 SUMMARY

The computer is able to process, store, and manage many different types of media. Hypermedia and interactive multimedia encourage readers to move from topic to topic rapidly and nonsequentially. Navigating this media landscape, referred to by some as Hyperspace, can be complicated. Topics, called *nodes*, are connected by electronic cross references called *links*. In these systems material is accessed on related topics by clicking a mouse pointer on words or images that would serve as gateways to related information. A hypermedia document is created by both the author and the reader. The author creates and places the links, and the reader decides which links to follow.

Multimedia and hypermedia, like graphic processing, are changing the way information is communicated. The tools are just beginning to be used in cartography, mostly for the creation of what are called *electronic atlases*. While the technology is still in its infancy, it is clear that the combination of interaction and multiple media will have a great influence on the distribution, presentation, and use of maps.

7.6 EXERCISES

1. View a HyperCard stack. How was interaction and a linking structure incorporated within the stack?

2. Digital movies and associated players are available on most newer microcomputers. Examine the player programs. What type of interactive control is available while viewing the movies?

3. HyperText is a random access approach to reading. Can you imagine a Hypermap? What type of hyper-elements could be incorporated within a Hypermap?

7.7 REFERENCES

APPLE COMPUTER, INC. (1993) *Inside Macintosh QuickTime*. Reading, MA: Addison-Wesley.

BLATTNER, M. M., SUMIKAWA, D. A. AND GREENBERG, R. M. (1992) "Communicating and Learning through Nonspeech Audio." In *Multimedia Interface Design in Education* (NATO ASI, Series F), New York: Springer-Verlag.

BLATTNER, M. M., SUMIKAWA, D. A. AND GREENBERG, R. M. (1989) "Earcons and Icons: Their Structure and Common Design Principles." *Human-Computer Interaction* 4, no.1: 11–14.

BROWN, P. (1986) "Viewing Documents on a Screen" In *CD-ROM: The New Papyrus*, edited by S. Lambert, and S. Ropiequet. Redmond, WA: Microsoft Press, 175–184.

BUSH, V. (1945) "As We May Think." *The Atlantic Monthly*, July, 101–108.

DRUCKER, D. L. AND MURIE, M. D. (1992) *QuickTime Handbook*. Carmel, IN: Hayden Publishing.

HONE, R. (1993) *QuickTime: Making Movies with your Macintosh*. Rocklin, CA: Prima Computer.

HORN, R. E. (1989) *Mapping Hypertext*. Lexington, MA: Lexington Institute.

KAY, D. C. AND LEVINE, J. R. (1992) *Graphics File Formats*. Blue Ridge Summit, PA: Windcrest.

MILLER, S. AND HARRIS, A. (1993) *The QuickTime How-to-Book*. San Francisco: SYBEX.

FURTHER READINGS:

BERK, E. AND DEVLIN, J. (eds.). (1991) *Hypertext / Hypermedia Handbook*. New York: McGraw-Hill.

BLATTNER, M. M. AND DANNENBERG, R. B. (eds.). (1992) *Multimedia Interface Design*. Reading, MA: Addison-Wesley.

FRAASE, M. (1990) *Macintosh Hypermedia: Vol. II, Uses and Implementations*. Glenview, IL: Scott, Foresman.

GAINES, B. AND VICKERS, J. N. (1988) "Design Considerations for Hypermedia Systems." *Hypermedia* 1, no. 2: 179–195.

HARRISON, S., MINNEMAN, S., STULZ, B., AND WEBER, K. (1990) "Video: A Design Medium." *ACM SIGCHI Bulletin* 21, no. 3: 86–90.

HOWELL, G. T. (1992) *Building Hypermedia Applications: A Software Development Guide*. New York: McGraw-Hill.

MARTIN, J. (1990) *Hyperdocuments and How to Create Them*. Englewood Cliffs, NJ: Prentice Hall.

NIELSEN, J. (1991) *Hypertext and Hypermedia*. Boston, MA: Academic Press.

WEYER, S. A. (1988) "As We May Learn." In *Interactive Multimedia: Visions of Multimedia for Developers, Educators and Information Providers*. Redmond, WA: Microsoft Press, pp. 87–91.

WOODHEAD, N. (1991) *Hypertext and Hypermedia: Theory and Applications*. Wilmslow, England: Sigma Press.

8

Computer Animation

8.1 INTRODUCTION

Animation usually brings to mind images of cartoon figures, particularly those created for children. Animations were at one time created manually, a frame at a time and transferred onto film. The computer has long been used to assist in the creation of individual frames, and is now being used for the interactive display of animations. Computer animations have become relatively common. Weather forecasters, for example, often use a computer animation to depict movement of weather systems or the flow of a jet stream.

The techniques associated with computer animation go well beyond those of film animation. Gersmehl (1990) distinguishes between seven types of computer animation that are applicable to cartography. These seven can be grouped into two categories: frame-based animation and cast-based animation. The two differ in how the animation is created. In frame-based animation the individual frames do not share common elements. Types of frame-based animation include the flip-book and the slide-show. With cast-based animation foreground objects can be made to move against a background. The more sophisticated approaches to cast-based animation allow background and foreground objects to move simultaneously. Examples of the cast-based approach include the sprite, stage and play, color cycling, metamorphosis (polymorphic tweening), and model-and-camera. This chapter reviews both of these approaches to animation.

8.2 FRAME-BASED ANIMATION

The frame-based approach is the simplest form of animation. The individual frames are created by a graphics or mapping program and then shown in quick succession. The illusion of movement or change is created by displaying the frames quickly but many frames are required for even a few seconds of animation.

The simplest frame-based animation is the "flip–book." This is a type of animation that is often created by children. It consists of a small stack of cards with a series of pictures showing the movement of an object, perhaps an animal or a person. The animation is created by flipping through the pages in rapid succession. The slide show is a second approach to frame-based animation. Essentially the same as the flip-book, this type of animation is accompanied by special effects between the frames, including fades, wipes, and dissolves.

The frame-based approach requires a series of individual frames. The material to be shown might be created by computer mapping, illustration, or image processing programs. The material could also be input through a scanner or video digitizer. A number of programs are available to assemble, store, and display this type of animation.

8.2.1 Presentation Programs

As described in the previous chapter, presentation programs are designed for the display of text and graphics. The programs essentially create a slide show, with each slide consisting of text, graphics, or a combination of the two. The graphics editor within the program can be used to create the individual frames of an animation. Pictures, sound, and digital movies may be also be incorporated into the presentation. Programs in this category include PowerPoint (Microsoft), Persuasion (Adobe), and Freelance Graphics (Lotus).

8.2.2 Card- or Page-based Programs

Card-based or Page-based multimedia authoring programs like HyperCard can also be used for animation. A flip-book animation would be created by simply flipping through a series of cards or pages. These programs incorporate visual effects such as wipes, zooms, and fades that can be added between cards to produce a slide-show type of animation.

The advantage of these programs is that they are designed for interactive use. The user can interactively select an animation and also control the speed of display. This type of interactive control of animation requires scripting. Chapter 9 includes HyperCard scripts for animation. Similar scripting structures are used by other programs in this category, including SuperCard (Allegiant) and Toolbook (Asymetrix).

8.2.3 Image Processing Programs

Image processing programs can also be used for animation. For example, NCSA Image, a public domain image processing and analysis program that reads and writes TIFF, PICT, PICS, and MacPaint formatted files. The program provides compatibility with many other applications, including programs for scanning, processing, editing, publishing, and analyzing images.

Animations are created by specifying the animation option and then opening a series of images. The Import and Export commands allow images with arbitrary binary and tab-delimited (spreadsheet) formats to be read and written. The program animates a stack by repeatedly displaying its slices (frames) in sequence. The animation speed is controlled by pressing keys 1 through 9 and the right and left arrow keys are used to single step through the slices. The "Oscillating Movies" option reverses the direction of animation at the beginning and end of a sequence.

8.2.4 Movie Editing Programs

Programs for editing digital movies combine graphic files, still images, and audio and video sequences. These programs usually include a set of filters for creating transitions and special effects. Programs in this category include Premiere (Adobe), MovieWorks (Interactive Solutions), and Passport Producer (Passport Designs). The latter two programs are also used for multimedia authoring.

Premiere can import a variety of graphic file formats and video and audio sequences. The construction window is used to create movies. This window consists of two video and three audio tracks, a transition track between the two video tracks, and a superimpose track (Figure 8.1). The project window contains the graphic, video, and audio material that is to be placed into the movie. An item is dragged with the mouse from the project window into the construction window and placed in either a video or audio track. The number of seconds that each frame is to be shown is defined by adjusting its length in the construction window. For example, the first map of the United States is shown for two seconds (see upper-left of construction window in Figure 8.1).

The "Special Effects" track adds a transition between the two video tracks. For example, a map can be placed in video track A and displayed for seconds 0 through 2. Another map, perhaps at a larger scale, can be placed in video track B and displayed between seconds 1 and 4. There is an overlap in the display of the two maps during second 2. In this overlap area the zoom special effect can be added to create the effect of "zooming in" on the second map.

Movie editing programs create movies that can be incorporated within other applications. For example, a movie can be placed within a text document and displayed by clicking on its icon. Programs for movie editing also compress the digital movies, helping to conserve disk space.

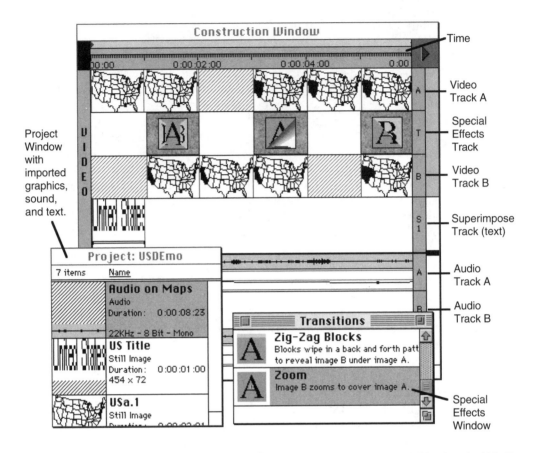

Figure 8.1   Frame-based animation with Adobe Premiere. Graphic material is placed within the construction window (Video Tracks A or B) and shown for a specific amount of time (time line is along the top of the construction window). Transitions can be added in the Special Effects Track and usually involve some type of fade or zoom from one graphic to another. (Courtesy of Adobe Systems, Inc.)

8.3 CAST-BASED ANIMATION

Cast-based animation is based on the concept of the *cel*. This form of animation is related to conventional film animation and the use of multiple transparent sheets to form a complete frame. A cel is an individual layer of a frame of animation, and a frame can be composed of many layers. A frame of an animation can consist of a background cel (a landscape) and a series of foreground cels containing an object that can be made to move on the background.

Moving objects in a cast-based animation using computer animation is done with a procedure called *tweening*, short for "in betweening." This procedure automatically creates a specified number of frames between two *key frames* with objects displaced proportionally in each frame. The technique makes it possible to create additional frames very quickly.

There are four types of two-dimensional cast-based animation: *sprite*, *stage and play*, *color cycling*, and *metamorphosis*. *Sprite* and *stage and play* are implementations of this foreground/background, cell-based concept in animation. The sprite is simply an object that moves against a static background. *Sprite* animation is commonly used in video games like PacMan. The *stage and play* type of animation is a more complex form of the sprite. Here the foreground object or actor can change in appearance or in its speed of movement. The background can also change simultaneously with changes in the foreground. More sophisticated video games (e.g., car racing) implement this type of animation.

Color cycling makes use of the way colors are represented on the screen of the computer. With this technique a path is defined and broken up into small segments. Each segment is given a slightly different color. Then the colors are shifted down each segment in the path to give the impression of movement (Figure 8.2a). This quick change in color along a path is used to simulate movement, as is commonly done with the representation of the jet stream on television weather maps.

Metamorphosis, also called polymorphic tweening, is the procedure used to change one shape into another (Figure 8.2b). For example, one could have a graphic object of a fish and another of a bird. One would then tell the software to transform the fish into the bird and it would calculate the "in between" frames that would make it appear as if the transformation were taking place. This can be combined with the sprite operation to show the fish jumping out of a lake, transforming into a bird, and flying into the air.

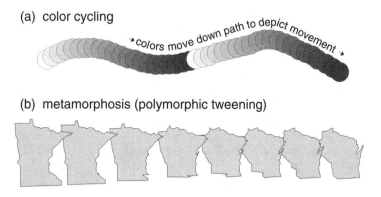

Figure 8.2   Two types of computer animation. Color cycling (a) simulates movement along a path by changing colors in zones along the path. Metamorphosis or polymorphic tweening (b) derives in-between frames between two given shapes. In the example shown here the outline for the state of Minnesota is converted into that of Wisconsin.

8.3.1 Programs for Two-Dimensional Cast-Based Animation

Two-dimensional computer animation programs are able to generate a series of "in–between" frames and work on a sub-frame, cel basis. The programs can import graphics files in a variety of formats and include an embedded graphics editor. Full-featured programs are usually fairly large and require a considerable amount of computer memory.

Programs with Limited Animation Effects

Programs in this category are not specifically designed for animation but implement one or two of the animation techniques described above. The most common effect is the "sprite," in which an object is moved across a static background.

SuperCard (Allegiant), for example, can move an object along a path. SuperCard has a HyperCard-like scripting language>. The script identifies the object to be moved and the point to which it will move. The object can be moved from one point to another, along the perimeter of a polygon or along a freehand curve. For example, the simulation of a car moving along a road is created by: (1) importing a graphic of a car and giving it the name Car, (2) drawing a line or polygonal roadway and naming it Road, and (3) attaching the following script to a button: Move graphic car to the points of graphic Road. The speed of the movement can also be controlled with the script and the frames can be recorded as a PICS animation file.

Full-Featured Animation Programs

Full-featured animation programs implement all or nearly all of the animation techniques previously described. The programs integrate a combined graphics-editing and scripting environment. Examples of programs in this category include: MacroMedia Director (Macintosh and Windows) AutoDesk Animator/ Animator Pro (DOS and Windows).

MacroMedia Director is an animation and multimedia authoring tool for presenting animated sequences with slides, video, and sound. Text, graphics, video stills, transition effects, and audio are combined for presentations. The program can create animated sequences and complete animated presentations that will run by themselves or from within a HyperCard stack (with MacroMedia player, a supplied utility program).

The program integrates a Painting window for creating artwork from scratch or modifying imported artwork, a Studio window for creating animations (called movies), and an Overview window for putting together simple animated sequences. The Overview window acts as a presentation tool, offering a control panel. The control panel consist of icons representing standard VCR controls, including play, rewind, pause, and the ability to dynamically change the tempo and the background color of the screen. The Studio window offers a variety of animation tools. It can display "cast members," which are elements of a presentation, such as pieces of text and graphics. The Studio provides access to the *score,* a table or spreadsheet containing cast members arranged in *frames.*

The *score* is used to define action in a movie (Figure 8.3). In the score the rows of the spreadsheet represent channels which are the elements on the stage where the

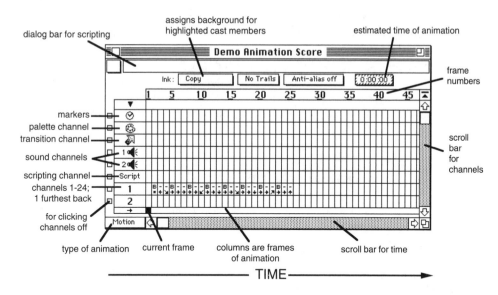

Figure 8.3   Score window from MacroMedia Director. The window is used to integrate multimedia material. (Courtesy of MacroMedia, Inc.)

animation is shown. The columns represent frames, a measure of time. The Studio window also provides access to the painting and drawing tools for creating or modifying cast members. This metaphor is used consistently throughout the program from the Overview window to the Studio score window. Cast members, including graphics, text, sound, color palette, can be lined up in each column and appear at the same time on the screen stage (Bove 1990, 182). The stage stores all elements: text, graphics, sound, color palettes, and individual film loops. The palette management controls and painting tools are designed to make color cycling simple to create (Bove 1990, 186). MacroMedia Director also includes an accelerator utility that converts Director movies into expanded versions that run faster and more smoothly, even when the movie contains very large and intricate color images (Bove 1990, 178).

AutoDesk Animator for DOS and Animator Pro for DOS and Windows are probably the most widely used animation programs. Animator is limited to a resolution of 320 x 200 with an 8-bit color look-up system. Animator Pro supports high-resolution graphics and a user selectable screen resolution (dependent on hardware restrictions). Animator Pro implements virtually all of the animation techniques, including cel-based animation, tweening, optical animation (swirling, twirling, spinning, flipping, squashing effect), and color cycling. Both programs have a paint module with built-in drawing and painting tools.

Animator and Animator Pro are supplied with a player program that allows the animations to be played back independently from the main program. The player is resolution independent and is capable of interactively branching to different animations or different parts of an animation. A built-in "C" programming language interpreter allows the users to create separate tools for the program. Both programs store animations in a standard FLC (pronounced "flic") file format.

Animator extensions are available for specific application areas. LANDCADD's VideoScapes works with Animator to create animations of photorealistic landscapes consisting of buildings, trees, cars, and people. Users can then interactively walk through the design.

8.3.2 Three-Dimensional Animation

The last form of animation is the *model and camera*. This is the most sophisticated of the animation procedures. With this technique one specifies both the position of the actor and camera, including the various lighting effects. Objects are defined with three-dimensional coordinates. Textures are added to the surfaces of the objects to make them appear more lifelike. This form of animation produces the most realistic results but is computationally intensive.

In 3-D graphic programs objects are modeled in the form of a wire frame consisting of polygons or small triangles. Three-dimensional wire frame objects can be created from two-dimensional objects through a *lathing* operation, analogous to a lathing machine used for woodworking. A side of the 3-D object is first drawn in two dimensions, then, by specifying *lathing,* the 2-D object is revolved in a circular fashion. The result is a wire frame of a 3-D object (Figure 8.4). Finally, a rendering procedure is applied by specifying lighting angle and intensity. The rendering step computes a shaded polygonal surface that corresponds to the lighting parameters.

Another method of creating a 3-D wire frame is to *extrude* a 2-D object into the third dimension. This involves duplicating the original object, offsetting it from the original, and then connecting the corresponding points (Figure 8.5). A rendering operation can then be performed.

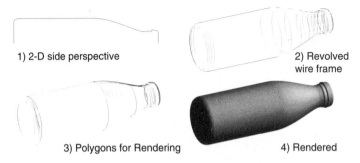

1) 2-D side perspective

2) Revolved wire frame

3) Polygons for Rendering

4) Rendered

Figure 8.4   Example of the lathing and rendering functions that are typical of 3-D graphics programs (e.g., Adobe Dimensions).

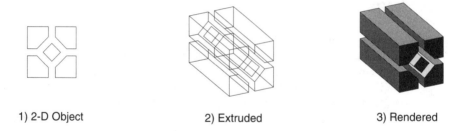

1) 2-D Object   2) Extruded   3) Rendered

Figure 8.5   Example of the extrude operation.   The procedure creates a three-dimensional figure from a two-dimensional object.

In the rendering step lighting effects such as reflection and shadows are computed for each polygon in the mesh. The three basic types of shading used in rendering, in increasing order of computational complexity, are: Gouraud, Phong, and Ray Tracing. Gouraud shading, used in Figures 8.4 and 8.5, renders only the effects of lighting. Phong shading adds shadows, reflections, and textures. Ray tracing calculates a shading value for each pixel or polygon based on the interaction of every ray of light in a scene, whether from a light source or a reflection. The texture and light reflectance properties of objects are also incorporated in ray tracing. The length of time required to render a scene, especially with ray tracing, is so great that distributed processing procedures have been developed. In distributed processing a series of computers work simultaneously on different parts of a task.

Although rendering has become standard in 3-D graphic programs the quality varies. Pixar's RenderMan is a highly regarded rendering program. This program lacks three-dimensional modeling capabilities but specializes in realistic three-dimensional rendering. Many 3-D programs output in the RIB (RenderMan Interface Bytestream) format so that this program can be used for rendering. RenderMan generates realistically rendered images in TIFF, PICT, EPS, or TGA formats.

Animation is supported by only a few 3-D graphics programs, and of these only a few are able to control every element of a scene over time, including lights and viewpoint — the true *model and camera* approach. Three-dimensional animation is an example of visualization, and borders on virtual reality. A major application has been in architecture, in which rooms can be created complete with furniture and then "walked through." Three-dimensional animation is also used for more general multimedia applications.

Programs for 3-D graphics that incorporate keyframe and morphing animation, and support QuickTime output are Crystal Topas (Crystal Graphics), Infini-D (Specular International), MacroMedia Three-D and MacroModel (MacroMedia), StrataVision 3D (Strata) and Playmation (Anjon & Associates). Other programs in this category include 3-D Studio (AutoDesk) and Studio Magic (Studio Magic Corp.).

8.4 SUMMARY

A variety of animation effects are possible with the computer. These can be grouped into two major categories — frame-based and cast-based. Frame-based animations such as the flip-book and slide-show can be implemented with a variety of programs, including presentation, multimedia authoring, and image processing programs. Cast-based animation requires the use of programs specifically designed for animation. The tools of cast-based animation include the sprite, color-cycling, and metamorphosis or polymorphic tweening.

8.5 EXERCISES

1. Evaluate a series of digital movies. How would you characterize the image quality? Is it possible to access any frame of the movie? Can you run the movie backwards?

2. Presentation programs (e.g., PowerPoint, Persuasion) can be used to implement an animation. Use a program in this category to create a brief cartographic animation.

3. Experiment with a three-dimensional modeling program. Extrude a map and then render the surfaces.

8.6 REFERENCES

Bove, T. and Rhodes, C. (1990) *Que's Macintosh Multimedia Handbook*, Carmel, IN: Que Corporation.

Gersmehl, P. J. (1990) "Choosing Tools: Nine Metaphors of Four-Dimensional Cartography." *Cartographic Perspectives*, no. 5: 3–17.

9

Scripting and Programming

9.1 INTRODUCTION

The computer is a medium of communication that is yet to be mastered. To master this new medium the computer user must be able to instruct the computer to do what is desired. Scripting and programming are ways of telling the computer to carry out specific operations, and allow the user to escape the limitations of existing programs. The development of more interactive and animated methods of mapping is dependent upon scripting and programming.

In the general sense any series of instructions executed by a computer is a *program*. The word *programming*, however, usually refers to the use of procedural languages such as Fortran, Pascal, and C. They are called procedural languages because the programs consist of a main program and a series of subroutines or functions, collectively called procedures. Programming with procedural languages requires a considerable amount of training. Much of this training involves learning the very specific syntax for the individual commands.

In contrast to programming, scripting languages rely on a more language-like syntax. Just as communication in language can be accomplished with different combinations of words, a script can use a different set of commands to accomplish the same task. For example, extra words can be inserted in the script for clarity. Generally, words like *the*, *is* and *to* can be included without affecting the operation of the script. Adding these words makes the script more "readable." Scripting languages are generally easier to learn than the standard procedural programming languages. This is especially true for the incorporation of user interface elements such as menus and buttons — generally the most difficult aspect of programming. Scripting languages generally have built-in structures that facilitate the incorporation of the user interface.

Procedural languages are also becoming more programmer friendly. The compilers can now better identify syntax errors, and debugging procedures help in determining logic errors. Some programming languages have optimization procedures that in effect rewrite a program to increase its execution speed. It is now much easier to program with these languages than it was during the first generation of microcomputers. An introduction to programming with procedural languages is provided in Appendix E. This chapter provides an overview of scripting, programming, and some basic user interface concepts.

9.2 HYPERTALK

HyperTalk is the scripting language associated with HyperCard. It is the best known of the first generation of scripting languages, and was included with all Macintosh computers for many years. It is designed to create an interactive environment in which the user can continually make decisions about the course of events.

HyperTalk scripting is relatively easy to learn, and the basic structure of the language has been followed in other multimedia programs (e.g., Allegiant SuperCard, Macromedia Director, and Asymetrix Toolbook). HyperCard is based on visual objects such as buttons that can be selected with a click of a mouse. This button is associated with a script that controls how objects respond after being selected by the mouse or the keyboard. All objects can be associated with a script.

HyperTalk can support both simple and complex tasks. For example, "buttons" may be interactively attached to commands that allow the user to move through a stack in a specific way. More complex applications of HyperTalk include address books, appointments, interaction with laser discs, and the incorporation of digitized sound. The program is most commonly used for interactive learning in education.

9.2.1 Stacks and Buttons

HyperCard files are referred to as stacks. A stack consists of a collection of cards that can also be associated with a script. A special stack called *Home* has an index function that can take the user to other available stacks (Apple 1989). HyperCard has five different user levels that define the extent of user control over the stack. These are:

     1. *Browsing* — the user can access the cards

     2. *Typing* — the user can type into fields

    3. *Painting* — the user can change the visual display

    4. *Authoring* — the user can make buttons linked to cards and stacks

    5. *Scripting* — the user has full control over creating objects and scripts.

The different levels allow the stack to be used by different types of users. The highest level is scripting.

    Figure 9.1 illustrates the steps involved in creating a new stack and using the Message command to change the user level to scripting. After a new stack has been created, a blank card is presented that represents the first card of the stack. Buttons, text, graphics, and sound can be easily added to this single card.

    A button is added by selecting "New Button" from the Objects menu (step 6, Figure 9.1). A field for text and a background object can also be selected. These objects can be moved any place on the card. A graphic can be added to the card by inserting a graphic file or *pasting in* a graphic from another program. A graphic illustration may also be created within HyperCard by using the bit-map graphics editor in the tools menu.

9.2.2 Sound

HyperCard can also access the sound generator within the computer. This is accomplished by first creating a button (Figure 9.2). Double-clicking on the button opens a dialog for changing the name of the button and adding a script. Selecting

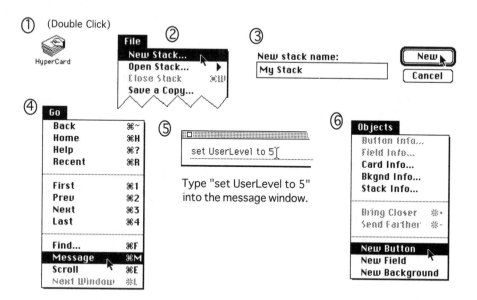

Figure 9.1   Creating a HyperCard stack. Double-clicking on the HyperCard icon opens the program. "New Stack" is then selected from the "File" menu. The stack is given a name in step 3. Steps 4 and 5 set the user level to scripting. Step 6 creates a button on the first card.

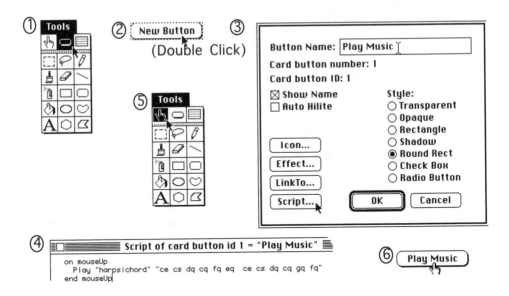

Figure 9.2   Assigning a script to a button in HyperCard. First, the button tool is selected from the Tools palette. Double-clicking on the button opens a dialog. The button can be named within this dialog (step 3). The script for the button is defined by clicking on the "Script" button. In step 4 the script is inserted between the "on MouseUp" and "end MouseUp" commands.   Then pointing mode control is reactivated in step 5. Finally, the button can be pressed and the script activated.

"Script" from this dialog makes it possible to enter the script for this button. The script consists of three lines:

```
on mouseUp
  play "harpsichord" "ce cs dq cq fq eq ce cs dq cq gq fq"
end mouseUp
```

The commands associated with the button specify that after the mouse is clicked (when the mouse is up again), the notes indicated will be played using the harpsichord sound built into the computer. The first letter of each two-character sequence specifies the note, and the second character denotes the duration of the note — *e* for an eighth note, *s* for a sixteenth, and *q* for a quarter note. The button becomes active after the script is saved, and the status is changed to "pointing" by clicking on the appropriate icon in the Tools menu.

Sound can be used in a variety of ways with maps: (1) a simple beep to verify a selection, (2) increasing or decreasing pitch to indicate scale change, (3) increasing static noise with greater data or map generalization, or (4) a spoken message accompanying a map to explain a distribution.

9.2.3 Transparent Buttons

The interactive map is based on the ability to select features by merely pointing to that feature on the map. This is accomplished in HyperCard by placing a transparent button behind the different features on the map. These buttons can be made to have a polygon shape, permitting the shape of a button to correspond to the shape of a feature. The illustration in Figure 9.3 depicts a part of a HyperCard stack. A map of a state by county is displayed by clicking on a state. The selection of a county presents a more detailed map of the county. The same procedure can be used to show street maps of increasing detail. A "zoom open" transition effect can be added to accentuate the effect of the zoom.

9.2.4 Animation with HyperCard

A frame-based animation can be created with HyperCard by cycling through a series of cards in quick succession. A script that displays a series of cards is presented in Script 9.1. This script would be attached to a button on the first card. After the button is selected, the "hide me" command hides the button. A "repeat" loop is then set up to first display card 1 and then the following nine cards. When the mouse is pressed, the "repeat" loop is ended. Card 1 is then displayed and the button is made visible.

There are two problems with the previous script. First, it will display the cards too quickly, theoretically at a speed of 60 per second. Second, there is no pause after the cards have been cycled through once. Script 9.2 adds a wait state both between individual cards ($3/60$ of a second) and after all cards have been displayed ($60/60$ of a second or one second). It also makes use of a variable called "counter" that is

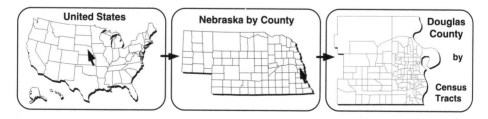

Figure 9.3   A cartographic zoom with a card-based multimedia program. Clicking on a state in the U.S. map displays the counties for that state. Clicking on a county presents a map of census tracts.

Script 9.1   A HyperTalk script attached to a button that displays a series of cards. After the button has been clicked, the button itself is hidden (hide me) and a loop is started that shows nine cards consecutively until the mouse is pressed. After this, the first card is shown and the button on it becomes visible again. (Script by Riley Jacobs.)

```
on mouseUp                                    ! after the button has been pressed
      hide me                                 ! hide the button
      repeat                                        ! start the loop called repeat
        if the mouse is down then exit repeat     ! if mouse pressed, exit loop
        go to card 1                              ! first go to card 1
        show 9 cards                          ! then show the following 9 cards
      end repeat                              ! end the repeat loop
      go to card 1                            ! return to card 1
      show me                                 ! show the button
end mouseUp                                    ! end procedure
```

incremented after each card is selected. In this way, a longer wait of "60" (one second) can be added at the end of each sequence of cards.

The HyperCard animation will be displayed more slowly during the first loop because the cards are being read from the disk for the first time. Beginning with the second loop, the display will be faster because the cards now reside in memory. It is possible to standardize the display rate by loading all the cards into memory before the animation is shown.

Script 9.2   A HyperTalk animation script that incorporates wait states. Similar to Script 9.1, this version adds a brief pause between each card ($3/60$ of a second) and after the set of cards have been displayed once ($60/60$ of a second or one second). The latter is accomplished with a counter variable that controls the repeat loop for the set of cards. (Script by Riley Jacobs.)

```
on mouseUp                                    ! after the button has been pressed
      hide me                                 ! hide the button
      repeat                                  ! start the loop called repeat
        if the mouse is down then exit repeat ! if mouse pressed, exit loop
        put 1 into counter                    ! initialize counter with the value of 1
        repeat until counter contains 10      ! begin second repeat loop
            go to card counter                ! go to the card number counter
            wait 3                            ! wait 3/60 of a second
            add 1 to counter                  ! increment the value of counter
        end repeat                            ! end loop that cycles cards once
        wait 60                               ! wait one second
        go to card 1                          ! display first card
      end repeat                              ! end the repeat loop
      go to card 1                            ! return to card 1
      show me                                 ! show the button
end mouseUp                                    ! end procedure
```

Figure 9.4 Typical dialog controls. The button accepts simple user responses of yes, no, or cancel. The check box is usually used to specify a choice from a list of possible options. Radio controls designate mutually exclusive options from a list of possible options. Static text is the unchanging text in a dialog, whereas editable text can be changed through keyboard input. The CNTL scroll is often associated with an editable text field for scrolling through a range of possible values. Icon and PICT are two graphic dialog elements that can be used to clarify choices. (Courtesy of Apple Computer, Inc.)

9.2.5 Dialogs

The most difficult aspect of programming is the user interface. In a point-and-click user interface, input is usually accomplished through a dialog. Essentially a dialog is a specialized window that contains different types of controls to set options. Most of these are simply selected with the mouse, although editable text dialogs accept input from the keyboard. Part of the purpose of the dialog is to limit the use of the keyboard — an error-prone input device. The different dialog controls are presented in Figure 9.4.

A dialog is created in HyperTalk with the "answer ... with" command. Script 9.3 creates a dialog with two buttons, in this case, "No" and "Yes." The conditional statements "if then" or "if then ... else" control the response to the various options.

HyperCard has extensive programming options, including the creation of graphics and data bases. The capabilities of HyperCard can be extended with so-called XCMDs

---

Script 9.3 Creating a dialog with HyperTalk. The "answer" command is followed by the text of the question. The "with" command then specifies the buttons that can be used to answer the question. The "if ... then...else...endif" sequence takes action based on the user input.

```
on mouseUp                              ! after the button has been pressed
   answer "This would be the question"  ! create a dialog with the static text

     with "No" or "Yes"                 ! add the two buttons for response
     if it is "Yes" then                ! if the "Yes," do the following

        Do these commands.

     else if it is "No" then            ! else if the "No," do the following

        Do these commands.

     endif                              ! end if conditional statement
end mouseUp                             ! end procedure
```

(x-commands). These are code segments or functions, written in a procedural language, that can be accessed through HyperCard. XCMDs are available for many different functions and often at a limited cost (e.g., Heizer Software).

HyperCard is a tool to integrate text, sound, and graphical materials of various types. It can be used to assemble interactive maps. Pictures can be used to depict the type of distribution that is being mapped. A video clip within a map might show a drive down a highway. The example stacks described here only represent a small portion of the potential for cartographic applications.

9.3 APPLESCRIPT

AppleScript is a scripting language that exchanges information and instructions between programs to automate repetitive tasks. For example, a script can transfer information from a data base to a graphing program, create a chart, and then insert the graphic into a text document. The language has sentence-like syntax similar to HyperTalk but contains all of the elements of procedural programming languages, including loops, conditional statements, arrays, and subroutines. AppleScript also allows users to create stand-alone applications.

The power of AppleScript is its ability to coordinate existing applications. For example, with a minimal amount of programming, one can tie together a data base, computer mapping, multimedia, and animation programs. AppleScript can move data into a computer mapping program, create a map, place the map in a multimedia program, and repeat the procedure to create an animation. This is a much easier alternative to writing a program that incorporates all of these functions.

A particularly attractive feature of AppleScript is its foreign language capability. Because it stores its code in a generalized internal format, AppleScript can translate its scripts into other languages. For example, a script written in Japanese can be used in the United States to perform the same function. Not only would the script perform the same function but it would appear in English to the user.

9.4 POSTSCRIPT

PostScript (Adobe) is a page description language that has become the industry standard for printing. Most printing is now done with printers that process PostScript codes. The printers contain a processor and relatively large amounts of memory to interpret the PostScript codes and convert them to a raster representation of 300 or more dots to the inch. Imagesetters also interpret PostScript and have resolutions up to 3,386 dpi. PostScript is also used for generating displays on screens as with a number of UNIX-based computers.

Chapter 5 discussed the importance of PostScript for illustration programs.

Portions of a map file in PostScript format were also listed (Figures 5.19 – 5.21). PostScript codes are generated whenever a file is printed to a PostScript-compatible printer. On the Macintosh these files can be viewed by choosing *PostScript File* as the destination in the print dialog. Rather than sending the file to the printer, this option will generate a text file with PostScript commands. The file can then be examined with a text processor. Utility programs can be used to send a PostScript file directly to the printer.

The PostScript language has been designed to produce graphic images and comes with a wealth of graphic operators. The basic graphic operators include:

newpath	initializes a new graphic object
closepath	closes the object
moveto	moves to the indicated position
lineto	draws a line to the indicated position
rmoveto	move relative to a current position
rlineto	draw relative to a current position
fill	shades the object
setgray	sets the gray level
setlinewidth	defines the line width of the current line

The use of these graphic operators is shown in Script 9.4.

PostScript also incorporates procedures that will group a set of operations together

---

Script 9.04   A PostScript script that creates the three overlapping boxes pictured below (Adobe 1985, 23).

```
newpath                 %Begin dark box
   252 324 moveto       %Move to this x,y coordinate
   0 72 rlineto         %Draw a line 72 units up from the current y position
   72 0 rlineto         %Draw a line 72 units over from the current x position
   0 -72 rlineto        %Draw a line 72 units down from the current y position
   closepath            %Draw a line to the first point
   .2 setgray           %Set the gray value to 80% ink
   fill                 %Fill the polygon with this shading
newpath                 %Begin middle gray box
   270 360 moveto
   0 72 rlineto
   72 0 rlineto
   0 -72 rlineto
   closepath
   .4 setgray
   fill
newpath                 %Begin lightest box
   288 396 moveto
   0 72 rlineto
   72 0 rlineto
   0 -72 rlineto
   closepath
   .6 setgray
   fill
showpage                %Send to printer
```

Script 9.5   A PostScript program using a procedure called box to create three overlapping boxes (Adobe 1985, 30).

```
% - - - - - - - - - -  Define box procedure - - - - - - - - - -
/box
   {0 72 rlineto
   72 0 rlineto
   0 -72 rlineto
   closepath } def

% - - - - - - - - - -  Begin Program - - - - - - - - - -
newpath                                    %First box
   252 324 moveto box
   .2 setgray fill
newpath                                    %Second box
   270 360 moveto box
   .4 setgray fill
newpath                                    %Third box
   288 396 moveto box
   .6 setgray fill
showpage                                   %Send to printer
```

---

under a single name or key, and store the information in a dictionary. When the key is used in a program, the associated set of operations is carried out. Script 9.5 uses a procedure called "box" to create the figure pictured in Script 9.4. This procedure is called after each of the three move commands. The procedural approach creates a program that is more compact, more readable, and can be more easily changed.

9.5 PROCEDURAL LANGUAGES

Most programs for computers are written with one of the procedural languages – C, Fortran, or Pascal. The major advantage of these programming languages compared to many of the scripting languages is their transportability. A program written on one computer can be made to work on another computer. The transporting of a program from one computer to another is not without problems. For example, there are differences in how the compiler writes to the screen or communicates with the mouse. The increasing similarity of operating systems (e.g., Macintosh and Windows) is making it easier to convert programs. Many of the programs that are described in this book have versions for multiple operating systems.

There are a number of different programming languages, as well as different implementations of the same language. Fortran was the most important language in the initial development of computer mapping, as well as computer graphics in general. Computer scientists had hoped that Fortran, first released in 1956, would be replaced by more structured and feature-laden programming languages. Fortran has continued to adapt, and in the latest release, Fortran 90, it incorporates most of the programming features found in the more modern languages (Metcalf and Reid 1990). Even in a window and menu environment, the Fortran programming language continues to be used because of its portability and fast execution speed.

C and Pascal are presently the major programming languages for microcomputers. These languages have the advantage of a pointer structure, with newer versions having object-oriented extensions. Object-oriented programming facilitates program development by identifying program segments as objects with special features. The C++ language is the most common object-oriented language.

It is important to point out that there is no best programming language. Each language has certain advantages and disadvantages. Many argue that C is the most functional programming language. Others point out that C is extremely cryptic, resulting in essentially illegible code. Fortran, it is said, is archaic and should be abolished, yet many of the advanced functions of Pascal and C have been incorporated in Fortran. Fortran also contains optimization features that leads to very efficient object code and the fastest execution times. Pascal is probably the easiest to program with the most "readable" code but its execution times are generally slower.

Programming knowledge is one of the criteria for choosing a language. Dershem and Jipping (1990, 35-36) list this along with six other criteria in choosing the best programming language — implementation, portability, syntax, semantics, programming environment, and model of computation. Often the selection of the best programming language is dependent upon the application to be programmed. Most programs in cartography now use C.

Programming the computer not only requires a knowledge of the programming language but also of the programming environment. On the Macintosh, this is the Macintosh Programmer's Workshop (MPW). MPW is a system for developing software that includes a compiler, an editor, command processor and development utilities. It facilitates the linking of code from different languages, the use of an object-oriented set of routines called MacApp, and developer products offered by other companies (West 1987).

The Windows Software Development Kit is the major programming environment for Windows. It includes the Windows application programming interface (API) that consists of functions, messages, data structures, and data types for the development of Windows applications. The API is a device-independent set of routines to access the variety of displays, printers, and other devices that are usually associated with Intel-based PC computers. Associated tools enable the programmer to design icons, dialog boxes, fonts, menus, and other elements of the user interface.

9.6 THE GRAPHICAL USER INTERFACE

Graphical user interfaces have become the standard way of interacting with computers. The use of graphical icons on traffic signs and for international events such as the Olympics is an established practice. As early as 1970 Easterby recognized the advantages of the international use of symbolic displays over those that are language-based. Research has shown that there are clear advantages to an iconic interface in computer programs. Lodding (1983) asserts that iconic interfaces can "reduce the learning curve in both time and effort, and facilitate user performance while reducing errors." He attributed these advantages to: (1) people's view of images as "natural," (2) the human mind's powerful image memory and processing capabilities, (3) the ease with which icons can be learned and recognized, and (4) images possessing "more universality than text." Gittens (1986) verifies the ease with which graphical attributes of icons such as style and color can be used to represent common properties of a collection of objects.

Graphical user interfaces have become standard on microcomputers and workstations since the introduction of the Macintosh. For the PC, Microsoft Windows and IBM's OS/2 are the two main window-based operating systems. Open Look, Motif, and X–Windows are two of the many graphic user interfaces that work within the UNIX operating system. The development of graphical user interfaces has led to some standardization in the use of icons. Figure 9.5 illustrates the common icons that are used in graphic design programs.

The graphic user interface is dependent on programs that are internal to the computer or operating system. The Macintosh stores many of these programs that are the basis of the user interface in the form of read-only memory or ROM (Apple 1992). With Windows the user interface is implemented with C run-time functions. These are special dynamic libraries that are linked with a program. In addition to implementing

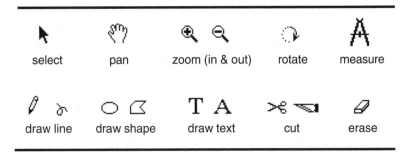

Figure 9.5   Icons used in graphic design programs. The icons shown here reflect the standardization of icons that has occurred in paint, draw, and illustration programs.

windows, pull-down menus, and dialogs, these routines handle all I/O events. An event includes all user actions including a mouse click, menu selection, and keyboard typing. Recognizing and processing events is central to the point-and-click interface.

Accessing the ROM procedures or Windows libraries and handling events is not a simple task. Experienced programmers have commented on the difficulty of programming computers with a point-and-click interface. It seems that as computers become easier to use, they become more difficult to program.

A variety of modular tools now exist to make it easier to incorporate a user interface and other programming tasks. MacApp (Apple Computer) accesses many of the user interface elements on the Macintosh. FaceWare is another modular package that assists in the programming of the user interface on the Macintosh. A number of function libraries are available for MS-Windows. Function libraries are available to assist with the user interface, graphic operations, manipulating images, and database operations, among other tasks.

9.6.1 Resources

To incorporate the user interface, programs are split into so-called "code" and "resource" forks. The code fork is written and compiled in a traditional computer language such as C, Pascal, or Fortran. If done properly, this part of the program does not contain any words or prompts that would be visible to the user. Resources are independent parts of a program that can be "edited" with a resource editing program. Fonts, menus, windows, and dialogs are all resources. The resource fork contains the definition of the menus, windows and dialogs, that is, all components of a program that deal with user interaction. Programming within a graphic user interface involves writing the code in a programming language and editing the resources in a resource editor. The resource editor on the Apple Macintosh is called ResEdit (see Alley and Strange 1991). The Windows resource editors are SDKPAINT (for icons, cursors and bitmaps), DIALOG (for dialog-box descriptions), and FONTEDIT (for creating font files).

ResEdit for the Macintosh opens and edits the resource fork of a file. ResEdit first displays a list of files in windows. Double-clicking on a file name opens another window with a list of all the resource types in the resource fork of the file. Double-clicking on a resource opens the resource, and it is here that the resource can be changed. There are a variety of resource types: MENU resources that define the menus, DLOG resources for the dialogs and WIND resources for the windows. These basic resources are found in any larger application. To examine these resources, initiate ResEdit and open an application (be sure not to save any changes).

An FCMD is also a resource much like a precompiled code library, having many different types of functions. A basic FCMD is the "primary event handler" that processes events returned by the GetNextEvent toolbox call that handles all user input. A "secondary event handler" FCMD is typically a "window-driving" module and is passed events directly from the primary event handler. Examples of a "secondary event handler" would be an editor or a graphics window that is a part of a program. A window can have a variety of embedded functions. A graphics window, for example, might be tied to a reduced-view window that depicts a smaller version of the graphic display.

9.6.2 Program Localization

Resources are the mechanism by which the user interface for a program is translated into another language. This process is called "Program Localization" (Apple 1991). A company interested in localizing a program will usually have a representative in a foreign country translate the menu items, window names, and dialogs with a resource editing program. The source code itself is left unchanged. Figure 9.6 depicts the "File" menu from the Apple Finder program in both English and German.

The localization process is becoming increasingly important as the international sales of microcomputers increase. Apple Computer, for example, earns more than half of its income from foreign sales. With less than 4.7 percent of the world's population, the United States will increasingly represent a smaller and smaller proportion of computer sales.

Localization is normally synonymous with translation. However, there are other aspects of the program that may need to be converted as well. For example, there are differences in the way numbers are formatted in different countries (Figure 9.7). In many European countries the use of the comma and period is reversed from what is

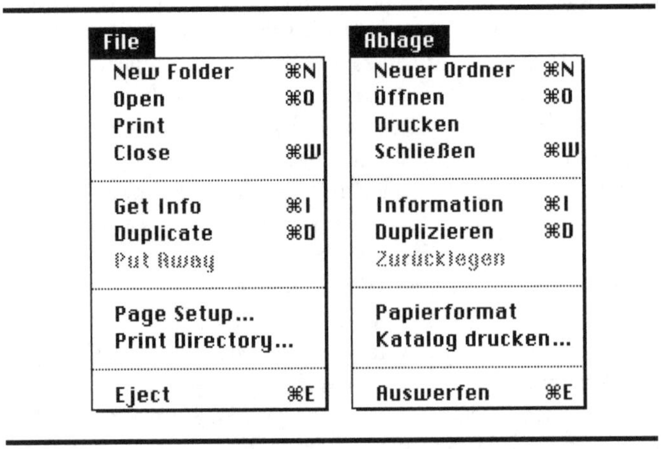

Figure 9.6   The File menu from the Apple Macintosh Finder (version 6.0.5) in English and German. Menu items are maintained in the resource file. The text associated with each menu item can be changed with a resource editor. (Courtesy of Apple Computer, Inc.)

Format	Country / Region
1 234,56	French, French Canada, Finland, Norway, Portugal, Sweden
1.234,56	Denmark, Netherlands, Flemish Belgium, Germany, Iceland, Italy, Spain, Yugoslavia
1 234.56	Greece
1'234.56	Switzerland (French and German)
1,234.56	Other countries including North America

Figure 9.7   Different number formats in different countries. The way numbers are written varies between parts of the world. These differences can be handled through the use of resources.

common in North America. These formatting differences are usually implemented within the resource fork as well so that the source code need not be altered.

Other culturally specific aspects of a program needing localization include: (1) the formatting of monetary values, (2) the use of the proper calendar, Gregorian or one of the lunar calendars, (3) the formatting of the times of the day, (4) the proper sorting of character expressions in different languages, and (5) the use of alternate keyboards. Specialized resources handle these localization tasks.

9.6.3 A Graphic Dialog

True graphic user interfaces make minimal use of text. A dialog can be made more "graphic" through the use of ICON and PICT resources (Figure 9.4). These resources store object-oriented or bit-map graphics. Because such graphic depictions need not exceed the resolution of the screen, and because bit-maps can be more quickly drawn on the screen, bit-map graphics are normally used to create the graphic user interface.

A graphic can be imported directly into ResEdit and associated with a button (Figure 9.8). Each element of the dialog is associated with a value. If an element in a dialog is chosen, a message is returned to the program. Here a dialog presents the option of drawing a square or an oval. The button next to the shape would be selected to choose that shape. It is also possible to place an invisible button behind each shape so that the user could simply click-on that shape.

9.7 DRAWING WITH THE TOOLBOX

All drawing operations on microcomputers are controlled by routines that are embedded in the hardware. With the Macintosh Toolbox the drawing of a rectangle can be accomplished with the toolbox call FRAMERECT, which has as its only argument an array of four elements containing the upper-left and lower-right coordinates of a

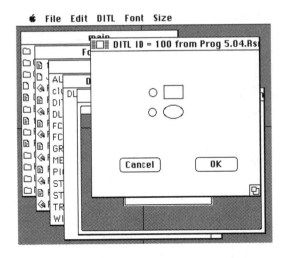

Figure 9.8   Defining a graphic "Shape Selection" dialog with radio control and PICT resources in ResEdit. A hidden button can also be placed behind each shape. In this case, the selection of the shape would take place by clicking directly on the figure and the radio control buttons would no longer be needed.

rectangle. Given an integer*2 array called *arect*, the call to FRAMERECT would be as follows:

```
call toolbx (FRAMERECT, arect)        (Fortran)
FRAMERECT (arect);                    (C)
```

Additional toolbox calls support other graphic options, including line width and color. A toolbox call can even be used to manipulate a rectangle array such as arect.

Toolbox procedures also exist that deal with polygons. As explained in Chapter 4, polygons can be defined in the computer as a series of x, y coordinates. Maps stored in this format can quickly be defined in the memory of the computer as a series of polygon objects, each referenced with a pointer (a number that defines the location of the polygon in memory). Once so defined, each polygon can be shaded according to a shading value and displayed on the screen. The pointer structure facilitates the coding:

```
do (i=1,nPolys)
      shadepoly (polys(i), grey(i))
      framepoly (polys(i))
repeat
```

where, nPolys is the number of polygons in the map, polys is an array of pointers to the individual polygons, grey is an array of shading values, and the routines *shadepoly* and *framepoly* shade the polygon and draw its outline, respectively.

9.8 SUMMARY

Scripting and programming are still in a process of evolution. New languages are being developed that are both easier to learn and more functional. Many complicated programming functions can now be done with easier-to-learn scripting languages. Even with the variety of application programs that are now available, it is still important to master some type of scripting or programming language to pursue specific tasks.

The most difficult aspect of both scripting and programming is the user interface. Handling the variety of prompts and user inputs can easily represent over 90 percent of the code for a computer program. To maintain consistency with other programs, it is important to use existing modules for the user interface. These modules may exist in hardware in the form of ROM or as part of the operating system software.

9.9 EXERCISES

1. Create a HyperCard stack with multiple cards that consist of text, graphics, and sound elements.

2. Examine one of the longer HyperTalk scripts within one of the HyperCard stacks supplied with the program. Try to write your own HyperCard script with a branching–control structure for the display of user-specified maps.

3. Use a resource editing program like ResEdit to change a menu item in a program. The menu items can be accessed through the menu resource. The title of a menu and the menu items can be altered in this resource.

9.10 REFERENCES

ADOBE SYSTEMS INC. (1985) *PostScript Language Tutorial and Cookbook.* Reading, MA: Addison-Wesley.

ALLEY, P. AND STRANGE, C. (1991) *ResEdit Complete.* Reading, MA: Addison-Wesley.

APPLE COMPUTER, INC. (1992) *Macintosh User Interface Guidelines*, Cupertino, CA: Apple Computer, Inc.

APPLE COMPUTER, INC. (1991) *Guide to Program Localization - A HyperCard Stack*, Cupertino, CA: Apple Computer, Inc.

APPLE COMPUTER, INC. (1989) *HyperCard Stack Design Guidelines.* Reading, MA: Addison-Wesley.

APPLE COMPUTER, INC. (1987) *Human Interface Guidelines: The Apple Desktop Interface.* Reading, MA: Addison-Wesley.

DERSHEM, H. L. AND JIPPING, M. J. (1990) *Programming Languages: Structures and Models.* Belmont, CA: Wadsworth Publishing.

LODDING, K. (1983) *Iconic Interfacing. IEEE Computer Graphics and Applications* 3, no. 2: 11–20.

METCALF, M. AND REID, J. (1990) *Fortran 90 Explained.* Oxford, England: Oxford University Press.

WEST, J. (1987) *Programming with Macintosh Programmer's Workshop.* Toronto: Bantam.

Part III

APPLYING THE TOOLS: EXAMPLES OF INTERACTIVE AND ANIMATED MAPS

The programs and programming languages discussed in the previous chapters can be used for interactive and animated mapping. The following two chapters describe examples of interactive and animated maps. Some of these examples are based on programs that are specifically designed for computer mapping whereas others involve the use of programs for multimedia and computer animation. The last chapter examines interaction and animation as the frontier of cartography.

10

The Interactive Map

10.1 INTRODUCTION

The interactive map consists of a map user, a computer, a pointing device, and a graphic user interface designed for the display and analysis of maps. Only a handful of programs in cartography effectively implement such a highly interactive form of map use. Many computer mapping and GIS programs still rely on the keyboard. In other programs the pointing device is simply used to select menu items.

The Great American History Machine (Miller 1988) incorporates many elements of the interactive map. Intended for the analysis of historical census data for the United States, the maps can be viewed side by side. The response time can be extremely slow, with a two minute wait sometimes required for maps to appear on the screen. Two other mapping systems that provide some measure of interaction with maps are based on Laserdiscs. The Domesday Project (Openshaw and Mounsey 1987) is a multimedia view of the United Kingdom with over 50,000 photographs. The Electronic Atlas of Canada (Siekierska 1984) incorporates a greater degree of interaction, permitting zooming, highlighting, and query operations.

The storage capacity of the CD-ROM has given rise to the development of a number of electronic atlases. CD-ROMs are the size of CD music discs and can be used to store about 600 MB of data. Map-based CD-ROMs mainly store existing maps and provide a viewer program to display the maps on a color monitor. CD-ROM products are available for the display of world maps, satellite images, topographic, and street maps and generally cost less than $100.

A different kind of map interaction with a map is made possible with a car navigation system in which the car's movements are reflected in a computer display on the screen. City tour guides combined with electronic atlases are changing how people use city maps. Finally, programs for computer mapping and GIS implement a number of interactive features. Atlas*GIS (Strategic Mapping Systems) allows the user to select a polygon and determine the name of the area and its value. MapInfo uses separate windows to display the map, the associated data, and graphs. The program also implements a "hot-view" link between the map and the data. Clicking on a map or data item will show the corresponding item in the other window.

Map interaction can be divided into three categories: (1) electronic atlases, (2) maps for navigation, and (3) data analysis. In this chapter we examine the three general types of map interaction.

10.2 INTERACTIVE ATLASES

Electronic atlases attempt to combine multimedia techniques with the display of maps. A few of these atlases implement hypermedia concepts with "hot spots" that can be selected and examined. CD-ROM atlases along with electronic encyclopedias have become very popular. The demand for electronic atlases is itself an indication of the need for more interactive methods of mapping.

Electronic atlases have yet to take full advantage of the medium's potential. For example, the interaction with maps may be limited to merely selecting different maps, analogous to turning a page in an atlas. Map animations are usually not included because they require specialized software for viewing. The atlases can be divided into three categories based on their content and purpose:

1. *Paper maps and pictures.* This category essentially consists of scanned paper maps and satellite images. To avoid copyright infringement, government maps and images are used as the source material. The scanned files are stored in a variety of graphic file formats, including TIFF and GIF. The scanned text on these maps is often illegible. These electronic atlases are usually supplied with an interactive program for the selection of the maps and images. In some cases this selection can be done by clicking on the features.

2. *Computer maps and figures.* This type of electronic atlas consists of maps that were specifically constructed for display on the computer screen. Because of the lower resolution of the computer screen, the maps do not display as many features. The text and colors have been selected for legibility. Facts and figures can be displayed by selecting a location on the map and hitting a button. Locations associated with an informational item, called hot spots, are indicated with a box or bold text. The information might consist of currency exchange rates for a country, population figures for states, or other demographic data.

3. *Electronic street atlases.* Electronic street atlases can be categorized by whether they use the Tiger data base or rely on another source for the street networks. Tiger is the data base compiled by the U.S. Census Bureau and supposedly contains

every street in the United States. However, as it is such a massive undertaking to maintain the files, the maps are rarely up-to-date, and while the streets are usually in the correct position relative to each other, the street segments themselves are not shown correctly. A straight street will appear to zig-zag back and forth between street intersections. In creating the data base, the Census Bureau was more concerned with coding all of the street intersections for address-matching purposes, not the form of the streets themselves. The TIGER data base does allow the creation of a seamless map within individual states. EtakMap (Etak) is an improvement over the TIGER files, adding "shape points" between street intersections to depict the streets more realistically. EtakMap also incorporates water features, new roads, and road classifications, including one-way streets.

A second type of electronic street atlas displays maps that were specifically designed for the computer screen rather than derived from the TIGER data base. These maps are more generalized and do not include all of the streets. These atlases may store the map in raster format so that the maps can be displayed more quickly. In some cases the maps contain advertisements for restaurants and hotels. Palm-top computers represent the major display medium for these maps.

10.3 PERSONAL NAVIGATION

Getting from place to place is one of the major uses of maps. Programs for way finding, called personal navigation, are another example of interactive mapping. Intended for use with lap-top or palm-top computers, this type of software finds the shortest distance between two locations and then displays the best route on a map. Compression techniques are used to store the digital map on a computer diskette. AutoMap for the PC (AutoMap), costing about $40, is an example of this type of software designed for interstate travel.

Other interactive maps combine an electronic atlas with route finding. For example, CityGuide (Axxis) integrates the interactive display of maps, route finding, and information on hotels and restaurants. The program can move to a requested address and zoom in or out. Landmarks can be interactively identified. The shortest distance between two points can be determined by clicking on the beginning and ending points, with the solution incorporating information on one-way streets. Options in route finding include the selection of local roads only, highways only, and the minimizing of zig-zag solutions. Hotel searches can be based on price, value, location, or proximity to a desired feature (e.g., convention center). Clicking a hotel name from a list of possible hotels presents a map showing the location of the selected hotel with a blinking icon. Clicking on the icon provides more detailed information on the hotel. Restaurants are handled in a similar way. The data base to support these tasks can be significant. The CityGuide for New York City requires 7 MB of disk storage. Smaller city guides are available for palm top computers.

Car navigation systems include a computer display of a road map and a moving symbol indicating the car's current position. The map displayed on the monitor changes as the vehicle moves, with the location of the vehicle updated through a dead-reckoning system. After the position of the vehicle is initially established, the system constantly monitors the movement of the vehicle (e.g., speed, right turns, left turns), but irregular movements can cause the system to lose its position.

A current development in car navigation uses global positioning system (GPS) receivers. These receivers allow the car's position to be constantly updated on the computer map without the use of dead-reckoning. GPS, created by the U.S. Department of Defense, consists of a series of twenty-four satellites that orbit the earth. Small receivers are used to communicate with these satellites and determine the current location in latitude and longitude. This information can then be integrated with a map display. The technology is already being used by companies like United Parcel Service (UPS) to monitor the location of their vehicles.

10.4 DATA ANALYSIS

The analysis of data is the search for patterns that help to explain the distribution and relationship between variables. The data might consist of income, education, or age of population. An interactive mapping system is a tool to aid in this research.

A card-based multimedia program (e.g., HyperCard) can be used to display a series of existing maps and allow the user to explore the relationship between the mapped distributions. For example, in portraying a distribution of the percent of births to mothers under the age of twenty in the United States, textual information can be presented along with the map to accentuate and explain the distribution. Figure 10.1 is an illustration of a HyperCard stack that represents a type of interactive map.

A series of such stacks could be organized around the topic of teenage pregnancy. The user would open up any one of the individual stacks and then be able to navigate through them. An individual stack would allow the user to select from a variety of options, including viewing the map, determining the state with the highest or lowest rates, viewing the five highest and lowest states, or transferring to another stack that depicts another related distribution.

10.4.1 Interaction with MacChoro II

Ideally the interactive map would be a product of a program for computer mapping. Unfortunately interaction in cartography is most often based on programs like HyperCard that are not specifically designed for the creation or the display of maps. Some computer mapping programs for the Macintosh, MS-Windows, and computer workstations incorporate a high degree of interaction in the display of maps. An example is the MacChoro II choropleth mapping program for the Apple Macintosh.

Creating the initial map with MacChoro II begins by selecting a spreadsheet

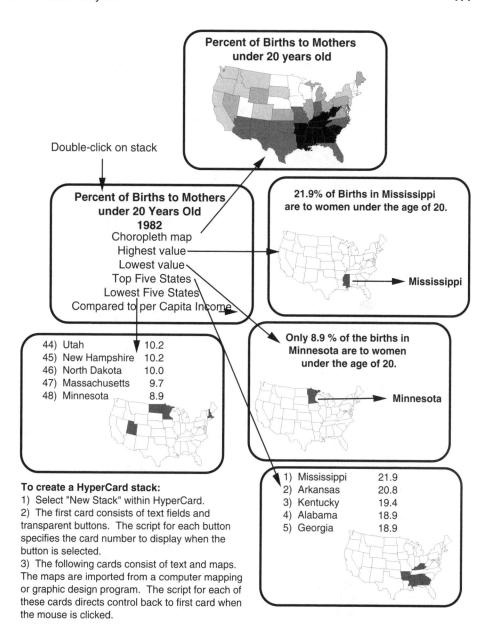

Figure 10.1   An example of a HyperCard stack for map interaction. An index card is used to point to the possible maps. The user clicks on a choice and the map is presented. This form of map interaction allows the user to explore a distribution more thoroughly than looking at a single map. The "link" between maps can be very involved.

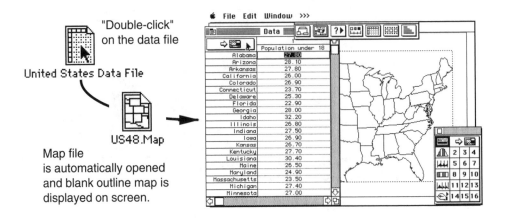

Figure 10.2   Opening a MacChoro II data file. Double-clicking on the data file automatically opens an associated map file. The map is placed in a graphics window as an outline map and the data is presented in a spreadsheet.

structured data file. The spreadsheet data file is "tied" to a specific map file. Opening a data file displays the associated map file as a blank outline map (Figure 10.2). Selecting a column of data and clicking on a draw map icon replaces the outline map with a shaded version of the outline map. Creating a map requires only three clicks of the mouse — selecting the data file, selecting a column from the data file and instructing the program to shade the map. A number of options are then available, including changing the default classification method and the number of classes.

A legend, neat line, and up to seven text elements may be added by defining a "bounding" rectangle and choosing the appropriate menu option. The legend can be displayed as a histogram, with the length of bars proportional to the number of data observations in each category. Four of the seven text elements are associated with the column of data, the fifth indicates the location of the map, and the remaining two describe the current classification (Figure 10.3).

The program incorporates four different options for the display of maps. These options are selectable with a dialog (Figure 10.4). The default option, *Descending*, displays the individual polygons from the top category to the bottom category. The next option displays the map by alternating between the high and low categories. For example, in displaying the map in Figure 10.3, all polygons in category 1 would be displayed first, then 4, then 2, then 3. This has the effect of accentuating the extreme categories. The *Group by Class* options groups the polygons of each class into a single graphic object. Clicking on any shading then activates all of the polygons with the same shading. The *Ascending by ID's* option displays polygons in ascending order by the identification number, usually arranged in alphabetical order.

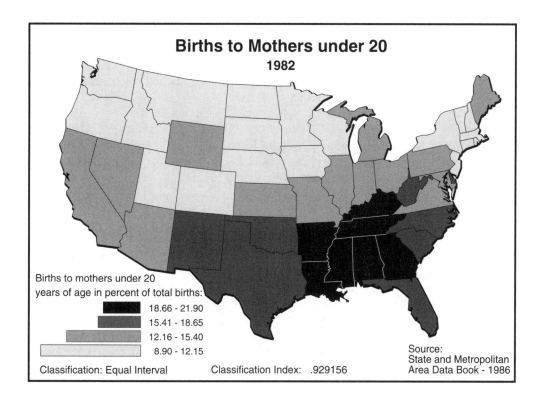

Figure 10.3 A choropleth map created with MacChoro II and edited with Adobe Illustrator. All map elements (map, legend, and text fields) are positioned with a bounding rectangle. The lengths of the legend bars are proportional to the number of observations in each category. The classification index is a relative measure of the quality of the data classification. A value of 1.0 indicates no classification error.

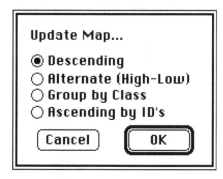

Figure 10.4 Map update options in MacChoro II. *Descending* displays polygons in descending order. *Alternate* first displays the highest category, then the lowest, then the second highest, and so on. *Group by Class* creates different graphics objects out of each grouping of polygons. *Ascending by ID's* displays the map from the lowest polygon ID value to the highest.

10.4.2 Interaction with EXPLOREMAP

EXPLOREMAP (Egbert and Slocum 1992) is a choropleth mapping program for MS-DOS developed at the University of Kansas. This program implements a variety of techniques for interactive mapping, effectively demonstrating the potential of a more interactive map form.

The program implements map sequencing in which portions of the map are presented in a particular order (Taylor 1991). The *Sequenced* option in EXPLOREMAP first displays the title of the map, then the legend title, and the legend boxes. After these map elements are displayed on the screen, the value range of the lowest class is shown in the legend, followed by the color in the legend box, and the corresponding distribution on the map. This is then followed by the next highest class.

This sequencing option is implemented in two separate ways. In the first a fixed delay of 1.5 seconds follows the placement of each map element. The second, an *Auto* option, allows the user to advance the display of the map by clicking on the mouse. It was noted through informal subject questioning that most users prefer the more interactive *Auto* approach.

The "Classes" option within the program permits the user to display any combination of classes. This is accomplished by allowing the user to select which classes are depicted by clicking on the corresponding legend box (Figure 10.5). Clicking on the legend box again causes that class to disappear from the map.

EXPLOREMAP includes a number of other functions, including bar charts, a display of the classification statistics, and a subset option to specify the desired range of data to be displayed.

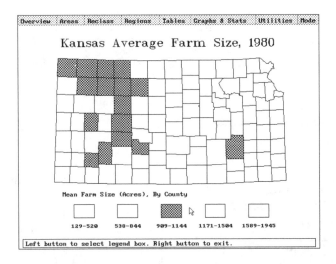

Figure 10.5   Classes option within EXPLOREMAP. Clicking on the legend box shades the corresponding areas. (Courtesy of S. Egbert and T. Slocum.)

10.4.3 ArcView® (ESRI)

A prime example of interactive mapping is ESRI's ArcView®, a data exploration and viewing program that works in conjunction with ARC/INFO® data files. The program, designed for MS-Windows, Macintosh and UNIX, implements an icon-based interface to create maps interactively. The map is constructed through the *table of contents* that lists the items, called *themes,* that can be displayed on the map. Clicking on a theme in the *table of contents* presents a map in a separate window (Figure 10.6). A legend editor dialog controls how the map elements are displayed.

A number of graphic operations are possible with the map. These are accessed through the icons above the map and control such aspects of the display as zooming, saving the map to a file and finding a particular address on the map (Figure 10.7).

The tools palette implements a number of query operations that access the data base. The tools can be used for such operations as selecting all areas that are located along a line or fall within a polygon or a circle (Figure 10.8). This can be used to identify properties that would be affected by the replacement of an electric line or are in the proximity of a natural gas leak. The zoom function in the Tool palette also allows the user to zoom to a particular location.

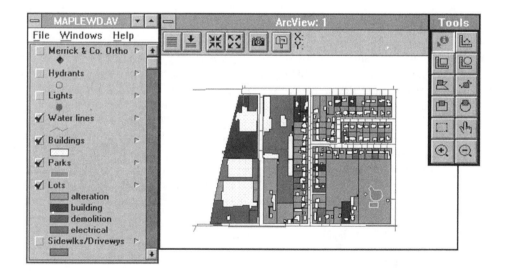

Figure 10.6   ArcView® implements an interactive mapping environment through a *table of contents.* Clicking on an item in the table displays the corresponding feature on the map. (Reprinted courtesy of Environmental Systems Research Institute.)

Figure 10.7   ArcView® Display menu. Buttons in the map display window perform operations specific to the currently displayed map. These include, from left to right, zoom to display all themes, zoom to the extent of selected theme(s), zoom-in, zoom-out, making a snapshot of the current display, and finding a specific address.   (Reprinted courtesy of Environmental Systems Research Institute.)

Figure 10.8   The Tools palette accesses the data base to perform a variety of query operations. The left side of the menu performs the following functions (top to bottom):   identifying a feature, measuring an area, selecting an area with a polygon, selecting an area with a bounding rectangle, zooming on the selected area, and zooming-in.   The right side of the menu: measure along a line, measure a radius, measure a line, select an area with a bounding circle, pan to position, and zoom-out.   (Reprinted courtesy of Environmental Systems Research Institute.)

10.5 SUMMARY

In the past cartographers have used the computer to assist in the process of bringing the map to paper. The computer was used to help the cartographer. Maps are now used directly from the computer screen. This introduces a new element to map use — interaction — and a new type of map — the interactive map. The interactive map is changing our conception of what maps are and how they can be used.

Interaction with maps can center on the graphic component of the map or on the underlying data. Interaction can be used for both general reference and thematic maps. Electronic atlases usually include both types of maps and implement different types of interaction. For navigation assistance the interactive map assists in route finding and the search for specific places. In data analysis the interactive map provides a mechanism to more closely examine quantitative data.

10.6 EXERCISES

1. Evaluate an electronic atlas. What forms of interaction are possible with the atlas? Could other forms of interaction have been incorporated?

2. Create an an interactive map for a socioeconomic variable, as in Figure 10.1, using a page-based or card-based multimedia program.

3. Evaluate a program for computer mapping or geographic information systems. List the different ways in which interaction has been incorporated in the program. How could the program be made more interactive?

10.7 REFERENCES

EGBERT, S. L. AND SLOCUM, T. A. (1992) "EXPLOREMAP: An Exploration System for Choropleth Maps." *Annals of the Association of American Geographers* 82, no. 2: 275–288.

MILLER, D. W. (1988) The Great American History Machine. *Academic Computing* 3, no. 3: 28–29, 43, 46–47, 50.

OPENSHAW, S. AND MOUNSEY, H. (1987) "Geographic Information Systems and the BBC's Domesday Interactive Videodisc." *International Journal of Geographical Information Systems* 1: 173–179.

SIEKIERSKA, E. (1984) Towards an Electronic Atlas. *Cartographica* 21: 2–3, 110–120.

TAYLOR, D. R. F. (ed.) (1991) *Geographic Information Systems: The Microcomputer in Modern Cartography.* Oxford, England; New York: Pergamon Press.

11

The Animated Map

11.1 INTRODUCTION

The animated map is a series of individual maps that are shown in quick succession for the purpose of depicting some type of trend or change. There are two major forms of the animated map: (1) temporal animations showing change over time and (2) non-temporal animations depicting a change caused by a variable other than time (e.g., deformation caused by a map projection). The animated map can also be categorized by different levels of interaction. Animated maps on film or video provide the least amount of interaction. Limited interaction is provided with digital video (e.g., QuickTime), in which individual frames can be accessed randomly and the user can control the speed and direction of display. With an interactive animated map the user directs the computer to create the individual frames of the animation and then controls their display. Few programs exist in cartography that implement this form of interactive animation. As a result, it is usually necessary to assemble an animation based on maps made in another program or use a program designed strictly for computer animation.

The primary purpose of the animation is to show a trend or a change that would not be visible if the maps were viewed individually. MacSpin (Abacus Concepts), a commonly used program for the analysis of three-dimensional data, clearly demonstrates the advantage of animation for the analysis of data. The program interactively "spins" three variables in an x, y, z coordinate space to depict configuration of data points (Figure 11.1). As the axis is spinned a "cloud" of points becomes visible. When the spinning is stopped, the cloud disappears. The program uses animation to reveal patterns that would otherwise remain hidden.

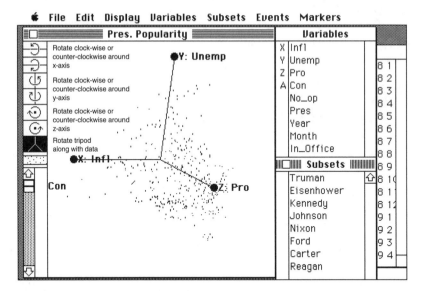

Figure 11.1   MacSpin data analysis program. The program animates the x, y, z coordinate system in three dimensions. Clusters of points that appear to be floating in space become visible while the animation is in progress. (Courtesy of Abacus Concepts.)

Animation has been applied to cartography in several ways. Chapter 3 reviewed examples of cartographic animations that have been done since the early 1960s. This chapter describes the creation of specific types of cartographic animation, including a program for interactive cartographic animation.

11.2 FRAME-BASED CARTOGRAPHIC ANIMATION

The simplest way to create a cartographic animation is by assembling a series of individual maps. These maps might be created by a program for computer mapping, scanned from existing paper maps, or input through a video source. Programs for the integration of animation sequences were presented in Chapter 9.

11.2.1 Movie Editing Programs

Programs for movie editing are able to combine video, audio, pictures, and graphics to create a digital movie. A cartographic animation can be created with a series of maps combined with other graphic or audio material placed within a digital movie. In the example explained here a "zoom" type animation is used to depict the location of Madras, India (Figure 11.2), using the Adobe Premiere movie editing program.

Three maps of India were scanned with a color scanner and saved as India Globe, India States, and India Madras. The maps were imported into Adobe Premiere and placed in the Project window. The first map, India Globe, was placed in video track A in the time slot between zero and two seconds (time indicators are given across the top of the construction window). The second map, India States, was placed in video track B in the time slot between one and four seconds. The third map, India Madras, is in Track A in the three to six second time slot. The FX track, between the two video tracks, is for the addition of special effects. Two special effects are used in the 1 to 2 and 3 to 4 second time slots. The first zoom effect has a small B inside the larger A, indicating that

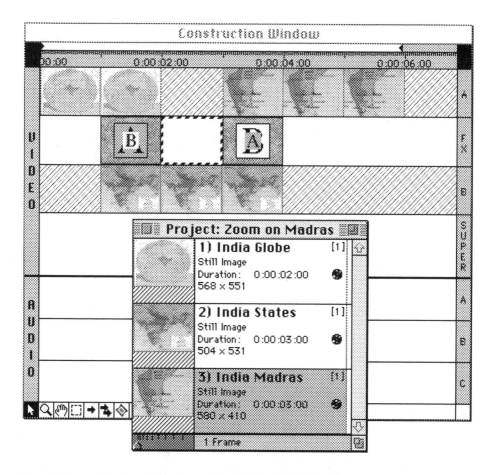

Figure 11.2   Creating a "zoom" animation with the Adobe Premiere movie editing program. A series of maps at different scales are scanned and assembled as an animation. The zoom visual effect is added between each scanned map. The B within the A transition indicates that the frame in track B is zoomed into while the A within the B zooms in on the frame in track A. The entire animation is six seconds long.

the image in the B video track will start as a small image and grow to cover all of the map in Track A. A similar "zoom" is implemented one second later, except here the map in track A grows to cover the map in track B. The entire animation requires six seconds.

Programs like Premiere can be used to create various types of cartographic animations, even including sound, but they are limited by their inability to create or modify maps.

11.2.2 Assembling a Satellite Image Animation

Satellite image animations showing the movement of cloud patterns are a common feature of weather forecasts. The individual images are taken by weather satellites that are in geosynchronous orbit at about 36,000 km above the earth. Individual images are transmitted to a receiving station on earth. These images are modified to include the state outlines and then distributed via satellite. The animations are created by placing a series of these weather images in a temporal sequence.

Weather satellite images can be obtained through the Internet. The Internet is a network of computers that can exchange files and messages (see Appendix B). Satellite images can be transferred from a remote to a local computer. The NCSA Telnet program can be used to establish a connection with the remote computer. Once logged on to the remote computer, a series of UNIX commands are used to change the directory, list the files, and retrieve the files from the remote computer. If your microcomputer is connected directly to a network, these files will then reside on the hard disk of your computer.

The weather images are in a GIF file format. Some programs that animate images can import these files directly (e.g., Adobe Premiere). For other programs the images may need to be reformatted. Shareware programs are available to convert the GIF files to TIFF format (e.g., GIFConverter). PhotoShop (Adobe) can also be used to reformat image files.

NCSA Image can be used to animate a series of TIFF files. This program, described in Chapter 6, is a public domain program designed for image manipulation. Image supports the organization and manipulation of a series of 2-D images as a 3-D array called a stack. A stack contains a set of related 2-D images, such as a movie loop or serial sections from a volume. The 2-D images that make up a stack are called slices. You can step through the slices using the ">" and "<" keys. The number of the current slice and the total number of slices are displayed in the title bar. The Stacks menu contains commands that work with stacks. *Windows to Stack* converts a set of 2-D images into a stack. *Animate* displays the images in a stack at a rate of up to 60 frames per second. Figure 11.3 depicts the individual steps involved for the opening each image, to specifying "Forward" from the Animation menu.

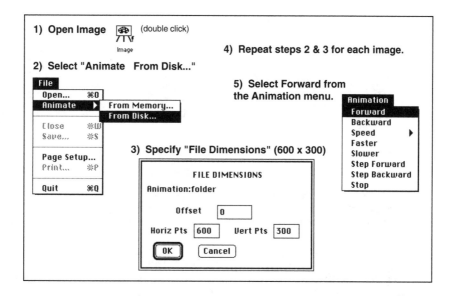

Figure 11.3   Creating an animation of weather images with NCSA Image. Each file is selected from the disk individually. When all files have been selected, the animation is started by selecting "Forward" or "Backward."

11.3 CREATING IN-BETWEEN FRAMES

Increasing the number of frames in an animation can have the effect of accentuating the change that is being depicted. Creating in-between frames, called *tweening*, is a common procedure in animation. Tweening, discussed in Chapter 8, involves the automatic creation of in-between frames to show the smooth transition in some characteristic of an object (e.g., shading value) or its shape. This type of graphic tweening can also be done with maps. For the animated map, however, we have the additional option of creating in-between frames by manipulating the data.

Interpolation is one method of manipulating data to increase the number of frames in an animation. The interpolation can be done either linearly or exponentially. In linear interpolation the difference between two known values is divided by the number of in-between periods. This number is then added to the base value incrementally for each intervening period. A spreadsheet program can be used to easily perform the calculations (Figure 11.4).

Linear interpolation assumes that the change between periods is a constant amount. For most phenomena this is not the case. Population, for example, increases in an exponential fashion. Figure 11.5 demonstrates the use of a spreadsheet and an exponential function to extrapolate future populations (the same formula can be used to interpolate between two known values, as above). The formula is based on the rate of population increase (natural increase divided by 100). Using this formula, a population

	A	B	C	D	E	F	G
1		Known	Computed	Computed	Computed	Computed	Known
2	States/Year	1970	1971	1972	1973	1974	1975
3	Alabama	35.10	36.16	37.22	38.28	39.34	40.40
4	Arizona	55.40	56.50	57.60	58.70	59.80	60.90
5	Arkansas	37.20	38.32	39.44	40.56	41.68	42.80
6	California	63.10	65.34	67.58	69.82	72.06	74.30
7	Colorado	59.50	62.28	65.06	67.84	70.62	73.40

Figure 11.4   Linear interpolation with a spreadsheet. Values are entered for the known years. A formula is then used to compute in-between values. The formula is entered in the top cell. The "Fill Down" command is then used to place a formula in all of the selected cells.

increase of 47.3 million is projected for the United States from 1993 to 2014. In that same period, Mexico, having a smaller base population, is calculated to increase by 55.9 million. The total populations or the population increases can be depicted with an animation, and the frames can be created on a yearly, even sub-yearly, basis.

Transition frames can also be created graphically. In Figure 11.6, the black areas in the frame on the left depicts land under the control of European settlers in 1784. The frame on the far right shows the same for 1810. The area lost by Native Americans in this time period is shown with the progressively darker shades of gray in the middle frames. Adding these in-between frames creates a smoother transition between the two time periods.

	A	B	C	D	E	F	G
1		Population	Nat. Inc.	Computed	Computed	Computed	Computed
2	Country/Year	1993		2000	2007	2014	2021
3	Canada	28.1	0.8	29.7	31.4	33.2	34.9
4	United States	258.3	0.8	273.2	288.9	305.6	320.6
5	Mexico	90.0	2.3	105.7	124.2	145.9	167.5

Figure 11.5   Computing future populations with a spreadsheet program. Population figures are given in millions. The formula that is used for the calculation is $P_t = P_0 e^{rt}$, where P_t is the future population, P_0 is the initial population, e is the base of the natural logarithms, r is the rate of natural increase (natural increase divided by 100), and t is the time period. EXP is an Excel (Microsoft) function that returns e raised to the power of a given number. e is the base of the natural logarithms, approximately 2.71828182845904. The formula can also be used to interpolate between two known values.

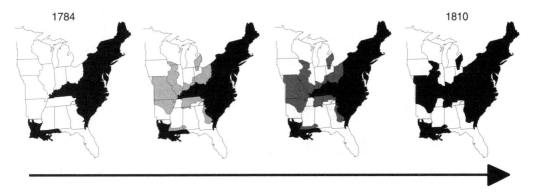

Figure 11.6   Creating in-between frames graphically. This series of frames shows the use of shadings to smooth the transition between two time periods. The maps show change in the areas controlled by European settlers from 1784 to 1810.

11.4 LEGENDS FOR THE ANIMATED MAP

Taking time to examine the legend is an important part of using any map on paper. It adds to the interpretation of the features and to the understanding of the distribution. In the animated map the speed of display is crucial. If the maps are shown too slowly, the communication of the change or trend is not achieved. There is no time to look at the legend on each map, and doing so would detract from the visual effect of the animation.

Attempts have been made to represent the passing of time in temporal animations. One technique is to display the year as a number that changes with each frame. But with this approach the user is not made aware of the position of the current frame within the overall animation. A second approach is to use a time scale (Figure 11.7). With a time scale, time is shown graphically so the user can see the relationship of the current frame to the beginning and ending frames.

There is still much to be learned about integrating legends with the display of animated maps. Any legend will detract from the animation because it moves the users attention away from the area where the maps are being displayed. It might be possible to integrate the legend within the map, perhaps as an individual frame. More interactive viewing environments would allow the users to stop the display and examine the legend.

11.5 ATLAS TOURING

Atlas touring, introduced by Monmonier (1990, 1992), encompasses a series of proposed techniques for the interactive use of electronic atlases. Most of these techniques use animation to help the user explore, or tour, geographic information. Atlas touring is based on two fundamental concepts — the *graphic script* and the *graphic phrase*.

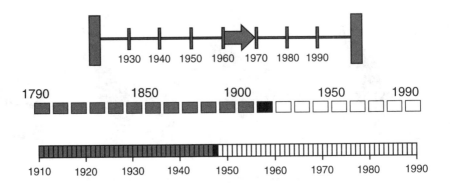

Figure 11.7   Three legends for a time series animation. The top legend uses a moving arrow to point to the current year. The bottom two legends use shadings. The gray area indicates the maps that have already been shown, the black area is the current map, and the white area indicates the maps yet to be shown.

The graphic script is a "meaningful sequence of maps, statistical diagrams, and other graphics generated or composed to explore or summarize salient trends and significant relationships that might be extracted from a geographic database" (Monmonier 1990, 5). It defines all aspects of the animation, including map content, symbolization, design, size, position, and duration of view. The graphic script may control multiple windows and specify orientation of view, tilt, completeness of the map, and scale (Figure 11.8a). A graphic script may also be used to represent a fly-through with "picture stations" along a flight path corresponding to particular points of view (Figure 11.8b).

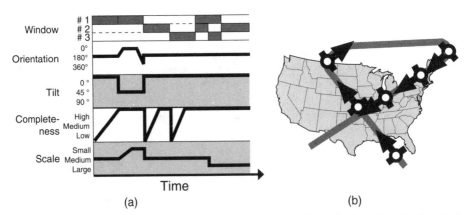

Figure 11.8   Illustration (a) depicts a multiple time-series graphic script that integrates the display of three variables in three separate windows. The script defines when the maps are being viewed (separately or at the same time) and changes in orientation, tilt, completeness of the map, and scale. Illustration (b) is a graphic script that defines the path of a fly-over with "picture stations" specified at particular points along the path.   (After Monmonier 1990, 52–53.)

A graphic phrase is the building block of the graphic script. It is a type of operator or instruction that is used for the development of scripts. As defined by Monmonier (1990, 21), a graphic phrase "produces a sequence of views that either focus on a particular face of a geographic distribution or set of distributions or address a specific analytical objective." It might define a sequence of views that zoom in on a particular region in a map. A toolbox of graphic phrases, available through a menu, could be used for developing a graphic script.

Atlas touring is a user interface for controlling the creation and viewing of cartographic animations. The graphic script and the graphic phrase are both mechanisms for user interaction with the computer. The touring concept suggests a drive through the countryside to view the scenery. Atlas touring specifies both the "drive" and the "view."

11.6 INTERACTIVE CARTOGRAPHIC ANIMATION

A major obstacle to the development of cartographic animation has been a lack of automated techniques for the interactive creation and viewing of map sequences. One of the few programs that implements an interactive cartographic animation is MacChoro II (Peterson 1993). Designed for the creation of choropleth maps, this program incorporates two types of cartographic animation. The first type of animation is related to the updating of single maps on the screen. The second is a flip-book cartographic animation of a series of choropleth maps at speeds of up to 60 frames a second.

11.6.1 Animation in Map Update

MacChoro II has two approaches to map viewing. Maps can be examined individually or as part of an animation. Individual maps are updated on the screen at the rate of between 20 to 50 polygons per second, depending upon the number of points in each polygon and the speed of the computer. The process of map updating is a form of cartographic animation.

Updating the map on the screen can be approached in different ways. For example, rather than updating the individual polygons in alphabetical order, as is normally done, a map could be displayed in a class sequential order (Figure 11.9). States belonging to the highest category would then be displayed first, followed by the next highest category and so on.

Although these map-update animations are only a few seconds long, they have a useful purpose. Because the classes are displayed individually, the eye focuses on the parts of the map that have a similar value. In this way the animation directs the attention of the viewer. There is no direct way to accomplish this with a static map.

Figure 11.9   The sequential display of a map with MacChoro II. The map can be placed on the screen by showing the states belonging to the highest category first. Then the states in the second highest category are displayed, and so on. The procedure requires from 1 to 3 seconds. This method of map updating is a form of animation.

11.6.2 Animation of a Map Series

MacChoro II also animates a series of maps. The user selects the classification method(s), the number of class(es), and the variable(s) to be animated. The individual maps of the animation are then created and placed in computer memory at a rate of about one map per second. After all maps reside in memory, they can be displayed on the screen at up to 60 frames per second.

Dialogs form the basis of the user interface. The animation is defined in the program by choosing the classification method(s) and the number of class(es) from one dialog and the variables to animate from a second dialog (Figure 11.10). A time-series animation can be created by selecting an entire series of data variables but only one classification method and class number. A classification animation (see Chapter 3) would depict only one variable and class number but all methods of data classification. A generalization animation would depict only one variable and data classification method but all numbers of classes. It is also possible to define a combination times-series, classification, and generalization animation. Once the particular animation has

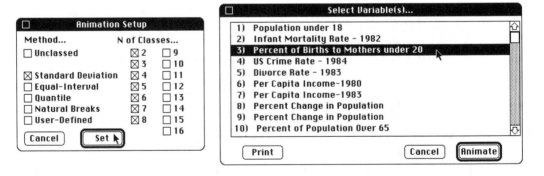

Figure 11.10   Dialogs used to create an animation with MacChoro II. The dialog on the left is used to specify the classification method or methods. In this case an animation is being specified with 2 to 8 classes using the standard deviation method of classification. The dialog on the right is used to select the data. In this case, only one variable is being chosen but multiple variables can also be selected.

been selected and the individual maps are created and placed in off-screen memory, the animation is shown as a continuous loop.

Interaction is brought into the animation through a pop-up palette menu that is accessible during the display of the animation (Figure 11.11a). The palette is used to change the display speed (up to 60 frames per second or as slow as one frame every 60 seconds), the direction (forward or backward) and for pausing and ending the display. It also provides access to a "manual animation" dialog (Figure 11.11b) that accesses the same options as the palette as well as a graphic display of the animation progress in the form of a circle. Each frame of the animation is represented by a sector in the circle, and clicking on a sector displays that particular frame. The range of an animation can also be changed by selecting the first and last frames in the control circle. The pop-up palette and manual animation dialog both facilitate an interaction with the animation display that is as not possible with most animations displayed with the computer.

11.6.3 Memory Requirements

The memory requirements for this form of cartographic animation are quite extensive. Depending on the size of the map display, each frame will usually require between 5 and 15 KB. For example, a 200 x 200 pixel map display would require 400,000 bits, 5,000 bytes (1 byte = 8 bits) or approximately 5 KB (1 KB = 1,024 bytes). A larger, more

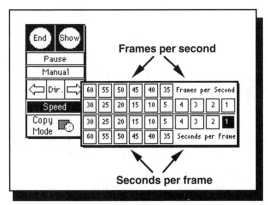

Animation Control Pop-up Palette

(a)

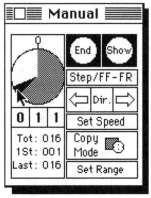

Manual Animation Dialog

(b)

Figure 11.11   Animation control dialogs. The pop-up palette dialog appears on the screen during the display of an animation in response to a click of the mouse. It includes basic controls that affect the display of the animation, including the ability to change direction and speed. Clicking on the speed option displays the "Frames per second/Seconds per frame" palette. The "Manual" dialog includes these basic controls along with a graphic display of the animation with a circle. Each segment of the circle indicates a frame of the animation. Individual frames of the animation can be accessed by clicking on a circle segment.

reasonably sized display of 300 x 400 pixels would require 120,000 bits or approximately 15 KB. A 255-frame animation at 15 KB per frame would need 3825 KB or 3.735 MB (1 MB = 1,024 KB). At least eight times this much memory would be required (30.6 MB) for maps with gray scale or color shadings. Image compression techniques such as those in Apple's QuickTime are becoming more common, and this would significantly reduce memory requirements, especially for the black and white maps used here. The animation could also be stored on a disk drive, but this would slow the animation creation and display processes.

11.6.4 Animation Display Speed

The purpose of an animation determines the speed at which it should be viewed. If the maps represent a particular sequence with only slight variation between them, a viewing rate of 60 frames per second is not excessive. To view differences between classification methods, a one second per frame rate seems more appropriate because the individual maps may be very different from each other. Certainly the viewing of animated maps is much different than viewing a map on paper. Although present map users are more accustomed to maps on a paper medium, future map users may not have this bias. The best approach, of course, is to allow the user to select the animation display speed.

11.7 SUMMARY

The concept of map animation has been a part of cartography for many years. It is only recently, however, that the technique has gained more widespread use, mainly because the tools to create cartographic animations are becoming more available. The microcomputer and the development of multimedia software have spurred the use of animation in cartography.

There are, however, few tools to create cartographic animations directly. Usually one must use programs that were designed for other applications, such as HyperCard and Adobe Premiere. Animation programs like AutoDesk Animator and MacroMedia Director permit the animation of objects within maps but the process is still tedious and time-consuming. Programs still need to be developed that combine interaction and animation in the display of maps.

11.8 EXERCISES

1. Use HyperCard to create a time series cartographic animation. Each card in the stack should represent data from a different year. Use the script from Chapter 9.

2. Use a movie editing program such as Adobe Premiere to assemble a series of scanned maps. For example, to depict the location of Bombay, India, you would scan maps of

Asia, India, and the western side of India. Experiment with the display times for each frame. Should they be consistent or vary between beginning and ending frames?

3. Retrieve a series of weather images from the Internet and create a weather animation with NCSA Image.

4. Create a generalization animation with a series of choropleth maps. What is the benefit of this form of map display?

11.9 REFERENCES

MONMONIER, M. S. (1992) "Authoring Graphic Scripts: Experiences and Principles." *Cartography and Geographic Information Systems* 19, no. 4: 247–260, 272.

MONMONIER, M. S. (1990) "Atlas Touring: Concepts and Development Strategies for a Geographic Visualization Support System." CASE Center Technical Report No. 9011, New York State Science and Technology Foundation.

PETERSON, M. P. (1993) "Interactive Cartographic Animation." *Cartography and Geographic Information Systems* 20, no.1: 40–44.

12

Frontier

12.1 INTRODUCTION

Maps have served an important role in the opening of frontiers. Early explorers charted new territory on maps. Settlers relied on maps for the division of land. The role of maps in the exploration of the unknown is still evident in the search for resources. Maps are created to determine the possible locations of needed resources such as oil. The map will always be an important tool of science and exploration.

Cartography now faces its own frontier. The frontier is a new medium, a medium that goes beyond paper and the individual, static map to one that presents a more dynamic view of the world. A medium that changes how space and place are conceived. This new medium compels us to approach maps differently. They exist for user interaction. It should be possible to "go into a map," to change what is presented and how it is depicted. Map use should be an active process that helps to visualize the world around us.

Animation helps to demonstrate that individual maps are only a snap shot in time. One should ask: What was before? What will come after? What trends would be evident if the time element could be viewed as an animation? The individual map is a snap shot not only in time but also in terms of how the data is represented. What non-temporal trends would be evident if a map were viewed along with other related data sets (e.g., age distribution in a city)? Finally, the individual map is a snap shot in the choice of the representational forms that are used to depict the world. A series of maps with different symbols or data classifications can also constitute an animation.

The previous chapters have dealt with the underlying concepts and the basic tools of an interactive and animated cartography. In this last chapter four fundamental concepts are identified and examined more closely: (1) the computer as medium, (2) the user interface, (3) the map as abstraction, and (4) creativity. Finally, the role an interactive and animated cartography in our own personal frontier is examined.

12.2 COMPUTER AS MEDIUM

Interactive and animated cartography is based on a conception of the computer as a medium of communication, not merely as a tool or instrument to store and manipulate data that ultimately will be printed as a map. The printed map cannot represent what is shown on the screen. It cannot capture the effect of interaction. It cannot be used for the display of an animation. Printing destroys the important associations in multimedia and hypermedia.

Each medium has its own set of biases that alter and control the form of information it transmits. One of McLuhan's arguments was that anyone who wishes to receive a message embedded in a medium must first internalize the medium so it can be "subtracted out" to leave the message behind (Kay 1990). Literate people have internalized the medium of print. How do we internalize the medium of the computer?

One answer may lie in the use of metaphors. The word metaphor is used here to describe a correspondence between what the computer does and how we should think about what it is doing. A metaphor relates our understanding of the computer to something with which we are already familiar. The "desk-top" metaphor, for example, has become the standard way or interacting with the computer. This metaphor is actually composed of multiple metaphors, including windows, folders, menus, and the trash can. These metaphors help us to work with the computer.

What metaphors can we use in cartography? The implicit metaphor for interaction is a conversation. The metaphor for animation is film. How do we extend these metaphors? How can we make the creation of maps more like a conversation? If we approach the computer with the use of metaphors — an emulation of a non-computer activity — the role of the computer is to refine and improve this activity, to extend it beyond what is possible without the computer. The way we refine and improve an activity performed by the computer and thereby internalize the medium is a function of the user interface.

12.3 THE USER INTERFACE

Creating an interface to the map user has always been a central concern of the cartographer. This concern is based on the need for user acceptance and communication. The computer medium simply presents another element in the map-user interface.

The computer-user interface is often simply thought of in terms of windows, pull-down menus, and dialogs. Of course, these are only the surface elements. Viewed in

these terms, the user interface is static. However, the user interface should be dynamic.
It should adapt to the user and stimulate their interests.

In user interface design much can be learned from computer games. The computer
game market is highly competitive. If the interface to a game is confusing, the player
simply abandons it. The user interface of a computer game must not only be functional
and easy to use, it must also be fun. The lessons to be learned from computer games are
(after Crawford 1990):

> 1. *Avoid the keyboard as the primary input device.* Many games and
> some programs operate entirely without the use of a keyboard. There
> are, of course, some applications where the input of text is required.
> Even here, however, the keyboard is often overused. Generally, its use
> can be limited to the naming of files.

> 2. *Place greater reliance on graphics and sound.* It is regrettable that
> programs for computer mapping rely so heavily on text. As with
> maps, graphics can express ideas directly. With text, communication
> is always indirect.

> 3. *Emphasize the intensity of the interaction.* The user needs fast
> responses from the computer. A delay of even a second is enough to
> break the user's concentration. Game designers speak of an
> "interaction circuit" in which the user and computer are in continuous
> communication. The user may opt to break the circuit but the
> computer cannot.

> 4. *Increase the level of complexity seamlessly.* The central problem in
> computer games is to move to increasing levels of complexity
> smoothly. Complexity is introduced by expanding the "vocabulary" of
> interaction. Vocabulary refers to the possible forms of interaction with
> the program. Expanding the possible forms of interaction clearly
> demonstrates that the user interface can be dynamic. It also shows that
> the user interface can constitute a learning environment.

The competitive environment of computer games has nourished some sound
concepts of user interface design. There is much that can be learned in the display of
maps and bringing the technology of interaction and animation to a wider audience.

In creating a user interface for maps we must also examine how maps are used. In
The Psychology of Everyday Things (Norman 1988) the author emphasizes that the tasks
are central, not the tools. He points out that a door has an interface — the doorknob and
other hardware. One should not have to think of using the interface to the door but rather
as simply going through the door. The interface is transparent. He suggests the
following priorities in user interface design (Norman 1990):

1. *The user.* What does the person need from the map?

2. *The task.* How can the purpose of map best be accomplished?

3. *Make the task dominate; make the tools invisible.* The user should not be overburdened by the interface.

4. *Do interaction correctly.* The user interface is dependent upon interaction. But, like a conversation, the interaction must conform to a set of possible responses. The response has to be expected to contribute to the overall user interface.

12.4 MAP AS ABSTRACTION

A map is an abstraction of the world, an intellectual transformation (Muehrcke and Muehrcke 1992). Some features are omitted while others are exaggerated, and in the process a map shows us more about reality than the examination of reality itself. According to Picasso, "Art is a lie that shows us the truth." If the truth is reality, it may be said that a "map is a lie that shows us the truth."

An analogy can be made between a map and a story. Like a map, a story can engage the human on a symbolic as well as an explicit level (Oren 1990). A story describes what is significant to the advancement of the plot and what deviates from the expected. An item is regarded as deviating from the expected when it does not fit into the context of the story. For example, the mention of a low-level detail raises its perceived importance in a story and may instigate a search for its relevance. The goal of the story is to engage the reader at both conscious and subconscious levels. Its purpose is to "tell more than it says."

A map is characterized by similar nuances. It shows us more than is there. It is this quality of maps that must be maintained in an interactive and animated environment. It may be that abstraction itself must be brought into interaction and animation. Indeed some of the methods of cartographic animation proposed in Chapter 3, including brushing, are forms of abstraction for cartographic animation.

12.5 CREATIVITY

Creativity is usually associated with the arts, not the sciences. However, it has a role in both. Creativity is needed when something is approached for the first time and there are few guidelines to follow. This is certainly the case with cartography and the new medium.

Where does creativity come from? How do we get new ideas? Is creativity a skill that can be learned or is it accessible to only a gifted few? There are numerous books on increasing creativity. Many stress the importance of making better use of the "neglected" right hemisphere of the brain to improve what is termed "visual thinking" (Edwards 1986). In general, these books challenge the comfortable, cautious, rational approach to problem-solving. New approaches to creative thinking can be worthwhile for individuals or groups who are attempting to find creative solutions to problems.

In a book entitled *A Kick in the Seat of the Pants* von Oech (1986) describes a method of thought provocation through role playing. He points out that we are often blinded by our own shortsighted view and tend to defend this view at all costs. He argues that role-playing can be both creative and liberating. von Oech describes four characters that are a part of each of us to varying degrees and can be used on a problem to different degrees:

> 1. *The Explorer.* This person gathers information. Activities include reading, asking others about their views, and deciding which issues need additional work or definition.

> 2. *The Artist.* This character generates new ideas. The most energetic and active; potential solutions and new problems are defined.

> 3. *The Judge.* This character evaluates and filters the ideas that have been generated. Some ideas are discarded. If the Judge has control in the beginning of the brainstorming process, creativity is short-circuited and new ideas do not emerge.

> 4. *The Warrior.* This character champions a particular idea and sets the course for the next round of problem-solving. This includes planning how the idea will be tested, evaluated, and developed.

None of these four characters can make an idea succeed without the participation of the other characters. Creative problem-solving can be enhanced by moving between the different roles. Knowing when to change roles is crucial. Mountford (1990) suggests a thirty-minute session for brainstorming, alternating roles between the four characters every seven minutes.

12.6 CONCLUSION: THE PERSONAL FRONTIER

The frontier is the boundary between the known and the unknown. As each of us comes to know the spatial world we live in, we push back the boundary of our own personal frontier. We do so by learning the streets of a city or the hills and valleys of a surrounding countryside. So it is too with the world beyond our own personal experience. Here we use maps to push back the frontier — to find out where things are and how things are related to each other. While we succeed in pushing back the frontier, it is always out there. There is always a point in our understanding of the world where the known ends and the unknown begins.

Every one of us is faced with our personal frontier. The movements we make in our immediate surroundings and the thoughts we have about the larger world are limited by this boundary. We all exist up to the boundaries of our own frontier. Maps can help us move beyond these boundaries. They can provide information and expand the boundaries of our own frontier. This is both the purpose and the promise of an interactive and animated cartography.

12.7 REFERENCES

CRAWFORD, C. (1990) "Lessons in Computer Game Design." *The Art of the Computer User Interface,* edited by Brenda Laurel. Reading, MA: Addison-Wesley.

EDWARDS, B. (1986) *Drawing on the Artist Within.* New York: Simon Shuster.

KAY, A. (1990) "User Interface: A Personal View." *The Art of the Computer User Interface* edited by Brenda Laurel. Reading, MA: Addison-Wesley.

MOUNTFORD, S. J. (1990) "Tools and Techniques for Creative Design." *The Art of the Computer User Interface,* edited by Brenda Laurel. Reading, MA: Addison-Wesley.

MUEHRCKE, P. C. AND MUEHRCKE, J. O. (1992) *Map Use: Reading, Analysis and Interpretation*, 3ed. Madison, WI: JP Publications.

NORMAN, D. (1990) "Why Interfaces Don't Work." *The Art of the Computer User Interface,* edited by Brenda Laurel. Reading, MA: Addison-Wesley.

NORMAN, D. (1988) *The Psychology of Everyday Things.* New York: Basic Books, 1988.

OREN, T. (1990) "Designing a New Medium." *The Art of the Computer User Interface,* edited by Brenda Laurel. Reading, MA: Addison-Wesley.

VON OECH, R. (1986) *A Kick in the Seat of the Pants.* New York: Harper Row.

WOOD, D. (1992) *The Power of Maps.* New York: Guilford Press.

Appendix A

Introduction to Programming

A.1 PROGRAMMING CONCEPTS

This appendix reviews the basic concepts of programming through a series of programs written in FORTRAN and C. FORTRAN has been chosen because of the use of this language in computer mapping, and C because of its importance in the programming of Macintosh, Windows, and UNIX systems. Introductory computer science courses often emphasize languages such as Pascal and Modula-2, and it is necessary to make the transition to a language that is more common in computer mapping applications. The examples are intended to include the basic tools available to the programmer.

A.1.1 Variable Types

Before examining the programs it is necessary to explain the different variable types, and the way they are represented in FORTRAN and C. Similar variables are used in the different programming languages. There are essentially three types, differing only in the number of bytes used in their representation: integer, real, and character.

An integer is a whole number and can be represented with varying combinations of bytes. FORTRAN handles 1-, 2-, 4- and 8-byte integers (specified as integer*1, integer*2, integer*4, and integer*8). C has three integer types — short, int, and long, although many C compilers have only two different types because both short and int are both 2 bytes (long is 4 bytes).

The number of bytes determines the magnitude of the number that can be represented. An integer*2, for example, is represented with 2 bytes or 16 bits (1 byte = 8 bits). A bit is either one or zero, that is, on or off. One of these sixteen bits indicates the sign of the number, either positive or negative. Therefore, 15 bits remain to represent the number. Each bit represents a certain number based on the power of 2. From right to left, these 15 numbers are 16384, 8192, 4096, 2048, 1024, 512, 256, 128, 64, 32, 16, 8, 4, 2, and 1. If all 15 bits are on, the number represented would be 32767 — the addition of all 15 numbers. If the bit representation is "000000000101011", the number would be 32 + 8 + 2 + 1, or 43.

A real variable specifies a number with a decimal. FORTRAN has 4-, 8- and 12-byte real numbers, specified with REAL*4 (default), REAL*8, and REAL*12. C has 4- and 8-byte reals, specified with FLOAT and DOUBLE, respectively. To print the value of a variable of type double or float using the printf command, we use the format descriptor %e for scientific notation (e.g., 88.26×10^{12}), or %f for regular decimal representation. FORTRAN has similar E and F descriptors.

In FORTRAN all variables between "i" and "n" in the alphabet are integers, unless otherwise defined. Variables represented by the remaining letters in the alphabet are reals. In C all variables must be specifically defined at the beginning of the program.

The character variable type handles text within a program. FORTRAN uses the CHARACTER statement to define a variable for text, whereas C uses the CHAR declaration. In both languages the length of the text must be defined. In FORTRAN a statement such as "CHARACTER*10 FirstName" defines a variable for 10 characters. The corresponding statement in C would be "CHAR FirstName[10]".

The integer, real, and character variables are the basic types of variables in all programming languages. Other variable types that are implemented in some programming languages include the BYTE, STRING, UNSIGNED, and VOID.

A.2 BASIC CONCEPTS THROUGH EXAMPLE PROGRAMS

A.2.1 Loop and Output

The loop is a basic tool that allows the repeated execution of a series of statements. Output in this case means that something is written to the screen. This program (see Program A.1) calculates the square roots of the numbers between 1 and 50. In FORTRAN, the loop begins with "do" and ends with the "repeat" statement. The write statement, within the do loop, writes the text within the single quotes and the variable "i" to the screen — specified with the output unit number "9" (this number is different between the different implementations of FORTRAN). FORTRAN contains a standard square root function, "SQRT." The square root function requires a real variable, so the value of "i" is assigned to the variable "x" before the square root function call. The pause statement interrupts program execution until the return key is hit.

Program A.1 FORTRAN Version

```
real*4 x                        ! define x as real with 4 bytes
do (i=1,50)        ! start the loop, with value of i from i to 50
  x=i  !assign the value of i to x (sqrt of integer not possible)
  x=sqrt(x)                          ! determine the square of x
  write(9,*) 'The square root of ', i,' is ', x! write it all out
repeat                                        ! repeat the loop
pause            ! wait for "return" before quitting to finder
stop
end
```

There are five major differences in the corresponding C version listed in Program A.2: (1) the stdio.h module within the ANSI library is needed for the "printf" function, (2) the "math.h" module, also within the ANSI library is needed to access the "sqrt" function, (3) the main() statement defines the beginning of the program; all statements within the {...} are in the body of the program, (4) the variable "i" is specifically defined as an integer, and (5) the "for" control statement is used to define the loop, and all statements within the second set of {...} are within the loop.

Program A.2 C Version

```
#include <stdio.h>        /* this library contains "printf" function */
#include <math.h>          /* this library contains "sqrt" function */
main ()
{
  int i;                 /* defines variable i as an int (2 bytes) */
  double x;                /* define variable x as a real (8 bytes) */
  for (i=1; i<= 50; i++)    /* beginning value; range and increment */
  {
   x=i;                              /* assign x the value of i */
   x=sqrt(x);       /* call function to compute the square root of x */
    printf("The square root of %d is %f\n",i,x);/* %f' prints a real */
  }
}
```

A.2.2 Reading Numbers into an Array

In this example program numbers from a file are read into an array, and the mean is calculated. The program will ask for the name of the file, open the file, and read the values. The CHARACTER (CHAR in C) statement defines the variable for the name of the file. File handling is the main form of input/output (I/O) and one of the most difficult aspects of programming. Serious errors can occur if data is not read properly. Error checking is written into this example to avoid such problems.

Two statements are introduced in the FORTRAN version of this program (Program A.3). The OPEN statement opens a file for use by the program. In this case the unit number "1" is associated with the existing "OLD" file given by DatFile. If an error occurs while it attempts to open the file, execution transfers to the statement labeled 1000 ("Unable to Open:"). The READ statement is also introduced here. The first READ uses an "a72" format to read the name of the data file. The second READ statement reads the numbers from the file defined with the unit number "1." When the end-of-file (EOF) is reached, execution is transferred to statement 500, and the sum of the values is written.

An array is a group of related variables under a single name. Individual variables within an array are referenced with numbers. For example, x(4) refers to the fourth element of the array x. Arrays make it possible to handle large amounts of data. Arrays are defined at the beginning of a program. To define an array the DIMENSION statement is used in FORTRAN although the REAL, INTEGER, and CHARACTER definitions are also acceptable (ex: REAL *4 x(1000) is equivalent to DIMENSION x(1000)). In C, square brackets are used after the name of the variable to specify it as an array (ex: float x[1000]). The starting point of an array differs between the two languages. Arrays in C are zero indexed meaning that the first element is zero. In FORTRAN the first element is 1. An array element referenced as x[0] in C is referenced as x(1) in FORTRAN.

The calculation of the mean is performed by summing the values using the variable called sum and then dividing this number by iCount. iCount is incremented by one with each pass through the loop. In the FORTRAN program, the "end-of-file" is determined with the READ statement option END=500. When the end-of-file has been found, iCount has already been incremented by one beyond the number of values in the file. For this reason iCount is decremented by one (iCount=iCount-1) after the loop has been completed.

Program A.3 FORTRAN Version

```
dimension Vals(1000)                 ! set aside 1000 spaces for Vals
character*72 DatFile    !define the variable to store name of file
write (9,*) 'Enter name of file as disk drive:file name '
write (9,*) "Ex: 'my disk drive:DATA.DAT'"
read(9, '(a72)') DatFile              !read the name of the data file
```

```
     open(unit=1,file=DatFile,status='OLD',ERR=1000)!error: go to 1000
     write(9,"('File successfully opened: ',a72)") DatFile
     sum=0.0                                         ! initialize sum
     iCount=1
     do
       read(1,*,END=500) Vals(iCount)   !now read values into the array
       write(9,*) 'The value read = ',Vals(iCount)
       sum=sum+Vals(iCount)
       iCount=iCount+1
     repeat
 500 write(9,*) 'The sum of the values in the file =  ',sum
     iCount=iCount-1               !iCount was incremented by one too many
     fMean=sum/iCount
     write(9,*) 'The mean of the values in the file =  ',fMean
     write(9,*) 'hit return'
     pause
     stop
1000 write(9,"('Unable to open: ',a72)") DatFile
     write(9,*) 'hit return'
     pause
     stop
     end
```

The C version of the program (Program A.4) introduces the FILE, SCANF, FOPEN, IF, EXIT, FSCANF statements, and function calls. FILE defines a pointer to a file. SCANF reads the name of the file (in percentage or string format) from the keyboard. FOPEN opens the file "DatFile" as a read-only ("r") format file. The IF statement checks if the file pointer "fin" is equal to NULL, indicating that the file was not properly opened. The EXIT statement stops program execution. FSCANF reads the values from the file "fin" with an "%f" or real format. Numbers are read until an EOF is encountered. The while statement is used to increment the loop. As a result, iCount is not incremented beyond the end of the file.

Program A.4 C Version

```c
#include <stdio.h>   /* this library contains the "printf" function */
#include <stdlib.h>    /* this library contains the "exit" function */
main ()
{
  float Vals[1000];      /* the array of real numbers to hold values */
  float sum,fMean;               /* reals variables for sum and mean */
  int iCount;                /* define the counter iCount as an integer*/
  char DatFile[72];
  FILE *fin;                               /* define input pointer */
```

```
printf ("Enter name of file as disk drive:filename\n");
printf ("Ex:  my disk drive:DATA.DAT\n\n");
scanf ("%s", DatFile);                      /* read name of data file */
fin = fopen (DatFile,"r");                      /* open data file */
if (fin == NULL)                    /* error in opening data file */
    {
     printf ("\nUnable to open:  %s!\n",DatFile);
     exit(1);
     }
    else                                    /* data file opened ok */
     printf ("File successfully opened:  %s!\n",DatFile);
iCount = 0;
sum = 0.0;
while (fscanf (fin,"%f",&Vals[iCount]) != EOF)
   {
  printf ("Value of x=%6.3f\n", Vals[iCount]);
    sum += Vals[iCount];                        /* sum the values */
    iCount=++;
  }
  printf("Sum of numbers in file=%6.3f\n",sum);
  fMean = sum / iCount;                        /* compute the mean */
  printf("Mean of numbers in file=%6.3f\n",fMean);  /* write it out */
}
```

A.2.3 Subroutines and Functions

Now that the values from the file are in an array a number of different calculations can
be performed. To do this another important element of programming is implemented —
the subroutine or function. Both of these allow the program to be divided into sections,
called subroutines or functions, each working on a specific part of the problem. The
method of programming using submodules is called top-down design. For example, a
main program with only one call to a subroutine could be written as follows:

```
call solve_all_the_problems_of_the_world
stop
end
```

The subroutine "solve_all_the_problems_of_the_world" could then call other
subroutines, which in turn call other routines that call other routines, that do the actual
work.

    FORTRAN distinguishes between subroutines and functions. Subroutines can
return several values, but a function can only return a single value. A C program is
defined as consisting of only functions, one of which must be named "main". A single
value is returned with a statement called "return". Notice in the C example

(Program A.6) that the function call to SD that calculates the standard deviation is made within the print statement. A subroutine call is used in the FORTRAN example (Program A.5) but a similar function call could have been made as well.

Another difference between the languages is in the methods by which variables are passed to subroutines and functions. In the FORTRAN program an entire new "Vals" array is defined in the SD subroutine (pass by value), thereby duplicating the storage. In the C program the function SD is defined as containing the array "Vals," making the duplicate array unnecessary (pass by reference).

In the FORTRAN main program the call to the subroutine SD is the only change needed to incorporate the subroutine from Program A.3. The call statement specifies the variables and arrays that are to be transferred to the subroutine, in this case the variables iCount, fMean, and the array Vals. In FORTRAN variables are passed into completely different variables in the subroutine. Therefore, the associated variable names in the subroutine can be the same or different. For example, the variable iCount in the main program is passed into the variable "n" in the subroutine.

Program A.5 FORTRAN Version

```
      dimension Vals(1000)              ! set aside 1000 spaces for Vals
      character*72 DatFile
      write (9,*) 'Enter name of file as disk drive:file name '
      write (9,*) "Ex:  'my disk drive:DATA.DAT'"
      read(9, '(a72)') DatFile
      open (unit=1,file=DatFile,status='OLD',ERR=1000)
      write(9,"('File successfully opened: ',a72)") DatFile
      sum=0.0
      iCount=1
      do
      read(1,*,END=500) Vals(iCount)    !now read values into the array
        write(9,*) 'The value read = ',Vals(iCount)
        sum=sum+Vals(iCount)
        iCount=iCount+1
      repeat
 500  write(9,*) 'The sum of the values in the file =  ',sum
      iCount=iCount-1              !iCount was incremented by one too many
      fMean=sum/iCount
      write(9,*) 'The mean of the values in the file =  ',fMean

      call SD(iCount,fMean,Vals)

      write(9,*) 'hit return'
      pause
      stop
1000  write(9,"('Unable to open: ',a72)") DatFile
      write(9,*) 'hit return'
      pause
```

```
      stop
      end
*
*  SD:   subroutine to calculate standard deviation
*
      subroutine SD(n,fMean,Vals)
      real*4 Vals(1000)          !create another array in this subroutine
      sumsq = 0.0
      do (i=1,n)                                          !start the loop
        diff = Vals(i) - fMean          !determine deviation from the mean
        sumsq = sumsq + diff**2.        !square deviation and add to sumsq
      repeat

!finish calculation of standard deviation
      SD = sqrt (sumsq / (n - 1))

!write out standard deviation
      write (9,*) 'Standard deviation is = ',SD
      return
      end
```

Adding a call to the subroutine or function is relatively easy. Only one statement, the "call" itself, is needed in FORTRAN. Two statements, the definition of the function and the call to it, are needed in "C." In the C program the function SD is defined in the main program as a function that returns a float value. The variables iCount and fMean, and the array Vals are defined as "belonging" to the array. The call to the SD function is made within the print statement.

Program A.6 C Version

```c
#include <stdio.h>   /* this library contains the "printf" function */
#include <stdlib.h> /* this library contains the "exit(1)" function */
main ()
{
  float Vals[1000];      /* the array of real numbers to hold values */
  float sum,fMean;                 /* reals variables for sum and mean */
  int iCount;             /* define the counter iCount as an integer*/
  char DatFile[71];     /* define DatFile at string with 72 elements */
  float SD (int iCount, float fMean, float Vals[1000]);
  FILE *fin;                                /* define input pointer */
  printf ("Enter name of file as disk drive:filename\n");
  printf ("Ex:  my disk drive:DATA.DAT\n\n");
  scanf ("%s", DatFile);                 /* read name of data file */
```

```
   fin = fopen (DatFile,"r");                          /* open data file */
   if (fin == NULL)                       /* error in opening data file */
      {
       printf ("\nUnable to open:  %s!\n",DatFile);
       exit(1);
       }
      else                                     /* data file opened ok */
       printf ("File successfully opened:  %s!\n",DatFile);
   iCount = 0;
   sum = 0.0;
   while (fscanf (fin,"%f",&Vals[iCount]) != EOF)
     {
      printf ("Value of x=%6.3f\n", Vals[iCount]);
       sum = sum + Vals[iCount];                     /* sum the values */
       iCount=iCount+1;
     }
     printf("Sum of numbers in file=%6.3f\n",sum);
     fMean = sum / iCount;                         /* compute the mean */
     printf("Mean of numbers in file=%6.3f\n",fMean);
     printf("Standard deviations=%6.3f\n",

     SD(iCount,fMean,Vals));
}
/***********************************************************************/
/*           SD: function to calculate standard deviation            */
/***********************************************************************/
float SD (int iCount, float fMean, float Vals[1000])
  {
     int i;
/* an integer counter */
     float diff,sumsq;/* variables to calculate deviations from mean */
     sumsq = 0.0;                                  /* initialize sumsq */
     for (i=0; i<iCount; i++)
       {
                                                    /* start the loop */
      diff = Vals[i] - fMean;                   /* calculate deviation */
      sumsq = sumsq + pow(diff,2.0);             /* square deviation */
       }
     return (sqrt (sumsq/(iCount - 1)));/* return the standard deviation
*/
  }
```

A.2.4 Sorting Numbers

The sorting of numbers is an ideal way to demonstrate some of the peculiar logic of programming. This example will also use a subroutine, but in the FORTRAN program values will be passed through a COMMON statement that defines variables and arrays used by both the main program and the subroutine (Program A.7).

The sort algorithm is referred to as a "bubble sort" because the lowest (or the highest) values "bubble" to the top. The sort algorithm here sorts from low to high. The algorithm uses a "do loop" that is embedded within another "do loop." The outer loop specifies one value in the array to which all other values in the inner loop are compared. If the second value is greater, the two values are switched. When the inner loop has been executed once, the lowest value in the array has bubbled to the top. The outer loop then proceeds to the second position in the array and the process continues. After the inner loop has executed the second time, the second lowest value is in position two. To change the sort from high to low, the if statement can simply be changed to "less than" (.lt.).

Program A.7 FORTRAN Version

```
      real*4 Vals(1000)              ! this sets aside 1000 spaces for Vals
      common/SortStuf/iCount,Vals(1000)!define common variables
      character*72 DatFile
      write (9,*) 'Enter name of file as disk drive:file name '
      write (9,*) "Ex:  'my disk drive:DATA.DAT'"
      read(9, '(a72)') DatFile
      open (unit=1,file=DatFile,status='OLD',ERR=1000)
      write(9,"('File successfully opened: ',a72)") DatFile
      sum=0.0
      iCount=1
      do
      read(1,*,END=500) Vals(iCount)     !now read values into the array
        write(9,*) 'The value read = ',Vals(iCount)
        sum=sum+Vals(iCount)
        iCount=iCount+1
      repeat
  500 write(9,*) 'The sum of the values in the file =  ',sum
      iCount=iCount-1              !iCount was incremented by one too many
      fMean=sum/iCount
      write(9,*) 'The mean of the values in the file =  ',fMean
      write(9,*) 'The number of the values in the file =  ',iCount
      call sort                    !call subroutine to sort the values
      write(9,*) 'The sorted values: '
      write(9,'(f6.3)') (Vals(i),i=1,iCount)          !write sorted list
      pause
      stop
 1000 write(9,"('Unable to open: ',a72)") DatFile
      write(9,*) 'hit return'
      pause
      stop
      end
*
*  sort:  subroutine to sort an array of numbers
*
```

```
subroutine sort
common /SortStuf/ iCount,Vals(1000)
n1=iCount-1  !outer loop will run to one short of the # of values
do (k=1,n1)                                !start the outer loop
    do (j=k+1,iCount)                      !start the inner loop
      if (Vals(k) .gt. Vals(j)) then
        save=Vals(k)   !if Vals(k+1)>Vals(k) then we interchange
        Vals(k)=Vals(j)!the two values by using the variable save
        Vals(j)=save                  !to temporarily  store Vals(k)
      endif
  repeat
repeat
return
end
```

In the C version (Program A.8) the Sort function is defined as "void" in the main program because it does not return a value. The variable iCount and the array Vals are defined as belonging to the Sort function. The definition of the function and the call to Sort are the only changes in the main program from Program A.6. The sorted values are printed out from within the Sort function.

Program A.8 C Version

```
#include <stdio.h>       /* this library contains the "printf" function */
#include <stdlib.h>      /* this library contains the "exit(1)" function */
main ()
{
  float Vals[1000];           /* defines array Vals with 1000 elements */
  float fMean;                    /* define variable to hold mean */
  int iCount;                          /* define iCount as int */
  char DatFile[71];      /* define DatFile at string with 72 elements */
  FILE *fin;                         /* define input pointer */

/* define void function for sort */
  void Sort;
  printf ("Enter file as disk drive:filename\n");        /* show how to */
  printf ("Ex:  my disk drive:DATA.DAT\n\n");
  scanf ("%s", DatFile);                        /* read name of data file */
  fin = fopen (DatFile,"r");                    /* open data file */
  if (fin == NULL)                        /* error in opening data file */
    {
    printf ("\nUnable to open:  %s!\n",DatFile);
    exit(1);
    }
    else                                  /* data file opened ok */
```

```
        printf ("File successfully opened:   %s!\n",DatFile);
  iCount = 0;
  sum = 0.0;
  while (fscanf (fin,"%f",&Vals[iCount]) != EOF)
    {
    printf ("Value of x=%6.3f\n", Vals[iCount]);
      sum = sum + Vals[iCount];                          /* sum the values */
      iCount=iCount+1;
    }
    Sort (iCount, Vals);                /* call function to sort values */
    }

                                        /* start of function to sort values */
void Sort (int iCount, float Vals[1000])
{   /* function is void because no value is returned to main function */
int j,k; float temp;
              /* outer loop will run to one short of the # of values */
    for (k=0; k<iCount-1; k++)                    /* start the outer loop */
    {
      for (j=k; j< iCount; j++)                   /* start the inner loop */
        {
        if (Vals[k] < Vals[j])                            /* ascending */
          {
          temp = Vals[k];  /* if Vals(k+1)>Vals(k) then we interchange */
          Vals[k]=Vals[j];     /* the two values by using the variable */
          Vals[j]=temp;          /* temp to temporarily  store Vals(k) */
          }
      }
  }
    for (j=0; j<= iCount-1; j++)     /* print the sorted list */
        printf("%6.3f\n",Vals[j]);
}
```

The programming concepts introduced here are the basis for all programming today. The only additional information needed would be methods of incorporating modules that have already been written. These modules can exist as C or Fortran libraries. For example, the C programs above use libraries called stdio.h, stdlib.h, math.h. These libraries contain a large number of different functions. The use of existing modules is especially important for the graphic user interface. On the Macintosh the modules that support the creation of windows, menus, and dialogs reside in a read-only memory (ROM) chip called the "toolbox." Microsoft Windows incorporates these routines within its operating system.

A.3 FURTHER READINGS

CLARKE, K. C. (1990) *Analytical and Computer Cartography*, Englewood Cliffs, N.J.: Prentice Hall.

DERSHEM, H. L. AND JIPPING, M. J. (1990) *Programming Languages: Structures and Models*, Belmont, CA: Wadsworth Publishing.

JOHNSONBAUGH, R. AND KALIN, M. (1989) *Applications Programming in C*, New York: MacMillan Publishing.

METCALF, M. AND REID, J. (1990) *Fortran 90 Explained.* Oxford, England: Oxford University Press.

PERRY, G. (1992) *C by Example.* Carmel, Ind.: Cue.

RIBAR, J. L. (1993) *Fortran Programming for Windows.* Berkeley, CA: Osborne McGraw Hill.

STROUSTRUP, B. (1991) *The C++ Programming Language.* Reading, MA: Addison-Wesley.

THINK C USER'S MANUAL. (1989) Cupertino, CA: Symantec Corporation.

WARD, T. & BROMHEAD, E. (1989) *Fortran and the Art of PC Programming.* Chichester, England: John Wiley.

Appendix B

Software Vendors

Abacus Concepts, Inc.
1918 Bonita Ave.
Berkeley, CA 94704
Phone: (800) 666-7828
FAX: (510) 540-0260
Products: MacSpin, StatView
Comments: MacSpin is a visualization program that spins an x, y, z coordinate system to
view three-dimensional data. StatView is a program for statistical analysis.

Adobe Systems, Inc.
P. O. Box 7900
Mountain View, CA 94039
Phone: (800) 833-6687
Customer Help: (800) 833-6687
Illustrator Help: (800) 986-6517
FAX: (415) 961-3769
Products: Illustrator (PC & Mac), Photoshop (PC & Mac), Premiere (Mac), SuperPaint,
Persuasion, PhotoStyler, IntelliDraw
Comments: Adobe products represent the entire spectrum of graphics-oriented software
from object-oriented Postscript graphics (Illustrator) to raster-oriented paint software
(Photoshop) and Quicktime-based presentation software (Premiere). Adobe recently

purchased Aldus Corporation. Former Aldus products include both graphics and
multimedia software. SuperPaint is a combined paint and object-oriented graphics
program. Persuasion is a multimedia presentation program. PhotoStyler is a color
editing and image processing program. IntelliDraw is a draw program. Most of these
programs are available for Macintosh and Windows. (FreeHand is now distributed by
Altsys Corp.)

ALL Systems Design, Inc
73 Lexington St.
Newton, MA 02166
Phone: (617) 630-0145
FAX: (617) 630-0116
Products: GeoVista
Comments: Produces choropleth maps with patterns or colors.

Allegiant Technologies Inc.
4660 La Jolla Drive, Suite 500
San Diego, CA 92122
Phone: (619) 535-4803
FAX: (619) 535-4890
Products: SuperCard
Comments: A multimedia product that implements a card structure and a scripting
language that is similar to HyperCard.

Altsys Corp.
269 W. Renner Pkwy.
Richardson, TX 75080-1343
Phone: (214) 680-2060
FAX: (214) 680-0537
Products: FreeHand
Comments: FreeHand is one of the major illustration programs for both the Macintosh
and Windows.

Anjon & Associates
714 E. Angelino St., Unit C
Burbank, CA 91510
Phone: (818) 566-8151
FAX: (818) 566-1036
Product: Will Vinton's Playmation
Comments: A 3-D graphics and animation program.

Apple Programmers Developers Association (APDA)
Apple Computer, Inc.
P. O. Box 319
Buffalo, New York 14207-0319
Phone: (800) 282-2732
FAX: (716) 871-6511
Products: Products to assist in scripting and programming for the Apple Macintosh.

Asymetrix
110-110th Avenue N.E., Suite 700
Bellevue, WA 98004
Phone: (800) 448-6543
FAX: (206) 637-1504
Products: ToolBook, Multimedia Toolbook, MediaBlitz, Compel
Comments: ToolBook, Multimedia Toolbook, and MediaBlitz are multimedia authoring
programs that integrate animation, full-motion video, and sound. Compel is used for
graphics presentations.

Autodesk, Inc.
2320 Marinshop Way
Sausalito, CA 94965
Phone: (800) 525-2763; (415) 332-2344
FAX: (415) 491-8305
Products: AutoCad, Autodesk Animator, Animator Pro
Comments: AutoCad is a standard CAD program for the PC, Macintosh, Sun, DEC and
HP workstations. Animator and Animator Pro are PC programs for the creation of
animations.

Automap, Inc.
1309 114th Ave., S. E., Suite 110
Bellevue, WA 98004
Phone: (206) 455-3552
FAX: (206) 455-3667
Products: Automap, Automap Europe
Comments: Automap includes detailed full-color maps that allow the user to access
51,921 cities and towns and over 359,220 miles of roads. Provides business,
commercial, and recreational travelers with maps, routing, points of interest, time and
distance, and educational features. Automap Europe provides similar travel information
for Europe.

Avian Systems, Inc.
1275 15th St., #15G
Fort Lee, NJ 07024-1929
Phone: (201) 224-2025
FAX: (201) 224-2566
Products: GAIA, FULLPIXELSEARCH, Vector That
Comments: GAIA is an image processing program for remote sensing applications.
FULLPIXELSEARCH is a pattern recognition program that works with TIFF and PICT
images. Vector That generates vector data from raster images.

Avid Technology, Inc.
One Metropolitan Park West
Tewksbury, MA 01876
Tel: (800) 949-2843
FAX: (508) 640-1336
Products: Media Composer, AudioVision, Media Suite Pro, Broadcast Products
Comments: Programs support digital media, both sound and video.

CACI
1100 N. Glebe Rd.
Arlington, VA 22201
Phone: (800) 292-2224
FAX: (703) 243-6272
Comments: Distributes demographic and business data.

Cartesia Software (formerly MicroMaps)
P. O. Box 757
Lambertville, NJ 08530
Phone: (800) 334-4291; (609) 397-1611
FAX: (609) 397-5724
Products: MapArt
Comments: MapArt includes graphic-oriented map files in Paint, Draw and EPS format.
The files work with both Macintosh and PC software.

Chadwyck-Healey
1101 King Street
Alexandria, VA 22314
Phone: (703) 683-4890
Products: SUPERMAP
Comments: A PC desktop mapping program that incorporates an interface to census data
on CD-ROM.

Claris Corp.
P.O. Box 58168
Santa Clara, CA 95052
Phone: (408) 727-8227
FAX: (800) 800-8954
Products: Brushstrokes, ClarisDraw, ClarisImpact, MacDraw
Comments: Paint, object-graphics and presentation programs for Macintosh and
Windows.

Claritas
53 Brown Rd.
Ithica, NY 14850
Phone: (800) 234-5973, (607) 257-5757
FAX: (607) 266-0425
Comments: Distributes demographic and business data.

Cognetics Corp.
P. O. Box 386
Princeton Junction, NJ 08550
Phone: (800) 229-8437
FAX: (609) 799-8555
Product: Hyperties
Comments: A hypertext application for the DOS operating system.

ComGrafix, Inc.
620 E Street
Clearwater, FL 34616
Phone: (800) 448-6277; (813) 443-6807
FAX: (813) 443-7585
Products: MapGrafix
Comments: A GIS program for the Apple Macintosh that supports map input.

COMPUneering Inc.
113 McCabe Crescent
Thornhill, Ontario L4J 2S6
Canada
Phone: (905) 738-4601
FAX: (905) 738-5207
Products: LANDesign, Data Collection Module, Export Module, LANDview,
Volumentrics Module, Import/Export Module
Comments: LANDesign is a surveying support program with COGO. LANDview
creates contour and three-dimensional maps.

Computer Systemics
806 Hill Wood Dr.
Austin, TX 78745
Phone: (512) 441-4583
FAX: (512) 443-6211
Products: GeoView
Comments: A topographic contour-mapping package limited to 500 data points per map.

Corel Systems Corp.
1600 Carling Ave., The Corel Bldg.
Ottawa, Ontario
Canada K1Z 8R7
Phone: (800) 836-7274; (613) 728-8200
FAX: (613) 728-9790
Products: CorelDRAW!
Comments: An object-oriented graphics program for Windows, UNIX and Macintosh.
The most widely used illustration program on PC computers.

Crystal Graphics
3110 Patrick Henry Dr.
Santa Clara, CA 95054
Phone: (800) 394-0700; (408) 496-6175;
FAX: (408) 496-0970
Product: Crystal TOPAS, Flying Fonts, 3D Designer
Comments: A 3-D graphics program that supports animation. Flying Fonts is for
animated text displays. 3D Designer supports modeling and rendering.

Decision Images
196 Tamarack Circle
Skillman, NJ 08558
Phone: (609) 683-0234
FAX: (609) 683-4068
Products: Roots and RootsPro
Comments: Desktop map digitizing for the PC and Macintosh.

DeLorme Mapping
P. O. Box 298
Lower Main St.
Freeport, ME 04032
Phone: (800) 227-1656; (207) 865-1234
FAX: (207) 865-9628
Products: Street Atlas USA
Comments: A CD-ROM and display program. Data base consists of all streets in the
United States.

Deneba Software
7400 S.W. 87th Ave.
Miami, Fla. 33173
Phone: (305) 596-5644
FAX: (305) 273-9069
Products: Canvas
Comments: A combined paint and draw program with numerous features (Macintosh
and Windows).

Digitial Wisdom, Inc.
Water Lane
P. O. Box 2070
Tappahannock, Virginia 22560-2070
Phone: (800) 800-8560
FAX: (804) 758-4512
Products: Mountain High Maps™
Comments: A variety of shaded relief maps of the world's continents and ocean floors
distributed on CD-ROM.

Discovery Systems International Inc.
7325 Oak Ridge Hwy, Suite 100
Knoxville, TN 37931
Phone: (615) 690-8829
FAX: (615) 690-2913
Products: Course Builder, Video Module
Comments: A multimedia authoring system for creating stand-alone applications. Video
Module is an add-on to Course Builder for control of a videodisc player.

Educational Multimedia Concepts, Ltd.
1313 Fifth Street S.E., Suite 202E
Minneapolis, MN 55414
Phone: (612) 379-3842
FAX: (612) 831-3167
Products: MacPresents, PCPresents
Comments: Multimedia authoring products for the Macintosh and PC.

Environmental Systems Research Institute (ESRI)
380 New York St.
Redlands, CA 92373
Phone: (909) 793-2853
FAX: (909) 793-5953
Products: Arc/Info, ArcView
Comments: Programs for geographic information systems. Most are designed for
workstations. ArcView is available for the PC, Macintosh and UNIX.

Equifax National Decision Systems
5375 Mira Sorrento Place, Suite 400
San Diego, CA 92121
Phone: (800) 866-6520
FAX: (619) 550-5800
Comments: Distributes demographic and business data.

ERDAS, Inc.
2801 Buford Highway, NE, Suite 300
Atlanta, GA 30329-2137
Phone: (404) 248-9000
FAX: (404) 248-9400
Products: ERDAS Imagine
Comments: A remote sensing and image-processing program for UNIX workstations
and Windows NT.

ETAK, Inc.
1430 O'Brien Drive
Menlo Park, CA 94025
Phone: (415) 328-3825
FAX: (415) 328-3148
Products: EtakMap
Comments: A digital road map data base. The maps generally depict urban areas more
accurately than TIGER files.

Evolution Computing
437 S. 48th Street, Suite 106
Tempe, AZ 85281
Phone: (602) 967-8633
FAX: (602) 968-4325
Products: EASYCAD, FASTCAD
Comments: Two inexpensive computer-aided design programs for the PC.

FaceWare
1310 N. Broadway
Urbana, IL 61801
Phone: (217) 328-5842
Products: FaceIt
Comments: A series of modules for the Macintosh that can be used with many different
programming languages to add a Macintosh style interface.

Fractal Design
P. O. Box 2380
Aptos, CA 95001
Phone: (408) 688-5300
FAX: (408) 688-8836
Products: Painter, PainterX2
Comments: Painter is a color paint program designed for artists. PainterX2 incorporates
objects and layering.

GeoQuery Corp.
387 Shuman Blvd., Suite 385E
Naperville, IL 60563-8453
Phone: (800) 541-0181; (708) 357-0535
FAX: (708) 717-4254
Product: GeoQuery
Comments: A mapping program specifically designed for business applications. It links
maps and data from existing sales databases.

Gold Disk Inc.
385 Van Ness Ave., Suite 110
Torrance, CA 90501
Phone: (310) 320-5080
FAX: (310) 320-0298
Product: Astound
Comments: A multimedia presentation program that handles both static presentations
and sound, animation, and QuickTime interactivity. A Macintosh product that includes a
player that allows finished applications to be played with Microsoft Windows.

Golden Software, Inc.
P. O. Box 281
Golden, Colorado 80402
Phone: (303) 279-1021
FAX: (303) 279-0909
Products: Surfer, MapViewer
Comments: Surfer is for contour and three-dimensional mapping. MapViewer creates
choropleth, graduated circle, three-dimensional prism, and point-symbol maps.

Graphsoft, Inc.
8370 Court Ave., Suite 202
Ellicott City, MD 21043
Phone: (410) 461-9488
FAX: (410) 461-9345

Products: Azimuth, MiniCad
Comments: Azimuth is a Macintosh program that produces maps of the world in one of nine standard cartographic projections. A perspective projection of part of the world is included as one of the projections. MiniCad is a CAD program for the Macintosh.

Great Wave Software
5353 Scotts Valley Drive
Scotts Valley, CA 95066
Phone: (408) 438-1990
FAX: (408) 438-7171
Product: ConcertWare Pro
Comments: A MIDI compatible software product that transcribes MIDI into editable files. Users can create a musical piece directly in notation mode.

Heizer Software
1941 Oak Park Blvd., Suite 30
P. O. Box 232019
Pleasant Hill, CA 94523
Phone: (800) 888-7667
FAX: (510) 943-6882
Products: HyperCard stacks and scripting software.
Comments: A variety of stacks and specialized scripts for HyperCard applications.

Highlighted Data Inc.
4350 N. Fairfax Drive, Suite 450
Arlington, VA 22203
Phone: (703) 516-9211
FAX: (703) 516-9216
Products: Electronic Map Cabinet
Comments: A CD-ROM product that contains street maps for cities in the U.S.

Infotec Development Inc.
3505-M Cadillac Avenue
Costa Mesa, CA 92626-1497
Phone: (800) 877-8796
FAX: (714) 549-8757
Products: LT4X and Delta3D
Comments: LT4X is for capturing and maintaining spatial data. Delta3D is a display program for digital elevation models.

Imsi
1938 Fourth Street
San Rafael, CA 94901
Phone: (415) 454-7101
FAX: (415) 454-8901
Products: TurboCad Designer
Comments: A CAD program for the DOS operating system.

Interactive Solutions Inc.
1720 S. Amphlett Blvd., Suite 219
San Mateo, CA 94402
Phone: (415) 377-0136
FAX: (415) 377-1964
Products: MovieWorks 1.1
Comments: A multimedia presentation tool that uses QuickTime to store and play
animations.

Intergraph Corp.
Huntsville, AL 35894-0014
Phone: (205) 730-8302
FAX: (205) 730-8300
Comment: A wide variety of hardware and software solutions for CAD and GIS.

Kansas Geological Survey
1930 Constant Ave.
Lawrence, KA 66047
Phone: (913) 864-5674; (800) 827-4844
FAX: (913) 864-5317
Products: Surface III
Comments: A contour and three-dimensional mapping program for the Macintosh.

LANDCADD International, Inc.
7388 S. Revere Pkwy., Bldg. 900, Ste. 901-902
Englewood, CO 80112-3942
Phone: (800) 876-LAND; (303) 799-3600
FAX: (303) 799-3696
Product: VideoScapes
Comments: A customization to Autodesk's Animator for the production of landscape
animations.

Lotus Development Corp.
55 Cambridge Pkwy.
Cambridge, MA 02142

Phone: (800) 343-5414
FAX: (617) 577-8500
Products: Freeland Graphics, Lotus 1-2-3
Comments: A presentation program for Windows. The Lotus 1-2-3 product is a
spreadsheet that has embedded mapping functions.

MacroMedia, Inc.
600 Townsend St., Suite 310W
San Francisco, CA 94103
Phone: (800) 288-4797
FAX: (415) 626-0554
Products: MacroMedia Director, MacroMedia Three-D
Comments: Multimedia programs that support animation. Three-D can create
animations in three dimensions.

MapInfo Corp.
One Global View
Troy, NY 12180
Phone: (800) 327-8627; (518) 285-6000
FAX: (518) 285-6060
Products: MapInfo
Comments: A desktop mapping program for market studies and land use planning.
Works under DOS, Mac, Windows, and UNIX.

MapWare
P. O. Box 50168
Long Beach, CA 90815
Products: The Map Collection
Comments: A PC program that creates a number of different types of maps including
contour, graduated circles, choropleth, block diagram, and map projections.

Media Cybernetics
8484 Georgia Ave., Suite 200
Silver Spring, MD 20910
Phone: (800) 992-4256; (301) 495-3305
FAX: (301) 495-5964
Products: HALO F/X
Comments: A grey scale image-editing software package. Windowed interface allows
access of up to ten images simultaneously.

MicroGrafx, Inc.
1303 Arapaho
Richardson, TX 75081
Phone: (800) 272-3729; (214) 234-1769
FAX: (214) 234-2410
Products: Windows Draw, Designer
Comments: Draw and illustration programs for Microsoft Windows.

Micro Map & CAD
9642 W. Virginia Circle
Lakewood, CO 80226
Phone: (303) 988-4940
Comment: A large number of map files in CAD format and translator program between different formats. The programs work with PC computers.

Microsoft Corporation
1 Microsoft Way
Redmond, WA 98052-6399
Phone: (206) 882-8080
FAX: (206) 936-7329
Products: MS-DOS, Windows, Windows NT, Paint for Windows
Comments: Microsoft is the world's largest software corporation. Their main product is the MS-DOS and Windows operating systems. The Paint program is a part of Windows.

NCSA Software Development — HyperCard
Scientific Animation Package
152 Computing Applications Bldg.
605 E. Springfield Ave.
Champaign, IL 61820
(FTP address: ncsa.uiuc.edu)
Product: HyperCard Animator, NCSA Telnet, NCSA Image
Comments: HyperCard Animator is a stack to aid in the development of HyperCard animations. Telnet is a communication and terminal emulation package. Image is an image processing and animation program.

Northeastern Digital Recording
2 Hidden Meadow Lane
Southborough, MA 01772-1700
Phone: (508) 481-9322
Service: Transferring files onto CD-ROM in Macintosh HFS and ISO 9660 formats.

NowWhat Software
2303 Sacramento
San Francisco, CA 94115
Phone: (800) 322-1954; (415) 885-1689
FAX: (415) 885-4092
Product: Small Blue Planet: The Electronic Satellite Atlas
Comments: A CD-ROM with satellite images and maps of the earth.

Objectic Systems, Inc.
P. O. Box 58292
Renton, WA 98058
Phone: (800) 859-9543; (206) 271-0204
Products: Fast Pitch Pro, version 2.1
Comments: A multimedia integration tool that works with HyperCard 2.1. Designed specifically for multimedia classroom presentations. The product integrates text, graphics, sound, and color/gray scale pictures, including scanned Photo CD images, QuickTime movies and videodisc material.

Paracomp Inc.
1725 Montgomery Street, 2nd floor
San Francisco, CA 94111
Phone: (415) 956-4091
FAX: (415) 956-9525
Products: Swivel 3D and Swivel 3D Professional
Comments: Macintosh programs for three-dimensional animation.

Passport Designs, Inc.
100 Stone Pine Road
Half Moon Bay, CA 94019
Phone: (415) 726-0280
FAX: (415) 726-2254
Products: Passport Producer
Comments: Integrates QuickTime movies, PICT images, PICS animations, text and audio for multimedia presentations.

Pixar
1001 W. Cutting Blvd.
Point Richmond, CA 94804
Phone: (800) 888-9856; (510) 236-4000
FAX: (510) 236-0388
Products: Typestry, RenderMan
Comments: Typestry converts Postscript and TrueType fonts into 3D images.
RenderMan specializes in three-dimensional rendering.

Rockware, Inc.
4241 Kipling St., Suite 595
Wheat Ridge, CO 80033
Phone: (303) 423-5645
FAX: (303) 423-6171
Products: MacGRIDZO
Comments: A program for contour and three-dimensional mapping for the Apple
Macintosh.

Scan/US
2032 Armacost Ave.
Los Angeles, CA 90025-6113
Phone: (800) 272-2687; (310) 820-1581
FAX: (310) 826-6863
Products: Scan/US
Comments: A demographic mapping program for Windows. The program displays
information on detailed raster maps.

ShapeWare
520 Pike St., Suite 1800
Seattle, WA 98101-4001
Phone: (206) 521-4500
FAX: (206) 521-4501
Product: Visio
Comments: An illustration program for Windows.

SIAL Geoscience Inc.
1400 Gouin Blvd. West
Montreal, Quebec
Canada H3M 1B1
Phone: (514) 339-2999
FAX: (514) 339-2997
Products: MacGoes II; MacGoes with Terneryplot
Comments: Programs for the analysis of satellite imagery.

Software Publishing Corporation
3165 Kifer Rd.
P. O. Box 54983
Santa Clara, CA 95056-0983
Phone: (408) 986-8000
FAX: (408) 980-0729
Products: Harvard Graphics, Harvard GeoGraphic
Comments: Harvard Graphics is an object graphics program. Harvard Geographics is
for choropleth mapping. Both programs are for the PC.

Specular International
233 N. Pleasant Street
P. O. Box 888
Amherst, MA 01004-0888
Phone: (413) 549-7600; (800) 433-7732
FAX: (413) 253-0540
Products: Infini-D
Comments: A three-dimensional design program that includes animation.

SPSS Inc.
444 North Michigan Avenue
Chicago, IL 60611-3962
Phone: (312) 329-2400
FAX: (312) 329-3668
Products: SPSS/PC+ Map, SYSTAT
Comments: SPSS/PC+ is a statistical analysis program that works in conjunction with
the MapInfo program. SYSTAT is a point-and-click statistical analysis program that
incorporates a number of mapping functions.

Strata, Inc.
2 W. St. George Blvd., Ancestor Square, Suite 2100
St. George, UT 84770
Phone: (800) 869-6855; (801) 628-5218;
FAX: (801) 628-9756
Product: StrataVision 3-D, StrataType
Comments: A 3-D animation program with rendering. StrataType is for rendering text.

Strategic Mapping, Inc.
3135 Kifer Road
Santa Clara, CA 95051
Phone: (408) 970-9600; (800) 866-2255 for demographic and marketing data
FAX: (408) 970-9999
Products: Atlas GIS for DOS and Windows, Atlas GIS for Windows Eurostat, Conquest
for Windows (specializing in data retrieval and mapping), AtlasView SDK software
developers kit (for programmer applications).
Comments: Desktop demographic mapping programs with varying levels of
sophistication. The company also distributes demographic and marketing data.

Studio Magic Corp. (formerly Brown-Wagh)
1680 Del Ave.
Campbell, CA 95008
Phone: (408) 378-3838
FAX: (408) 378-3577
Product: Studio Magic Personal Home Video Studeo
Comments: An inexpensive three-dimensional animation program for MS-DOS with
Windows.

Tactics International Inc.
16 Haverhill St.
Andover, MA 01810
Phone: (800) 927-7666; (508) 475-4475;
FAX: (508) 475-2136
Products: Tactician, Tactician with Buttons, Heavy Duty Tactician
Comments: Executive information systems and desktop mapping applications.

Terra Data Inc.
Bramblebush
Croton-on-Hudson, NY 10520
Phone: (212) 675-2971
FAX: (212) 675-2971
Products: Geocart
Comments: A map projections program for the Apple Macintosh that plots over 100
different projections. Maps can be exported in PICT or EPS formats.

TerraLogics Inc.
131 Daniel Webster Highway S., Suite 348
Nashua, NH 03060
Phone: (603) 889-1800
FAX: (603) 880-2022
Products: TerraView
Comments: A series of tools to analyze demographics and marketing-related statistics.

ThinkSpace Inc.
316 Cheapside St.
London, Ontario
Canada N6A 2A6
Phone: (519) 661-4006
FAX: (519) 661-3750
Products: Map II, HyperViewKit
Comments: Map II is a raster-based geographic information system. HyperViewKit is a
viewing program for raster images that works in conjunction with HyperCard.

USACERL (U.S. Army Construction Engineering Research Laboratory)
GRASS Information Center
P. O. Box 9005
Champaign, IL 61826-9005
Tel.: (217) 373-7220
Product: GRASS
Comments: A raster-based geographic information system for the UNIX operating system.

WordPerfect Corporation
1555 N. Technology Way
Orem, UT 84057
Phone: (800) 526-5059
FAX: (801) 228-5376
Products: DrawPerfect
Comments: An object graphics program for the PC.

Appendix C

Internet

Internet is an international computer network that links millions of academic, military, government, and commercial computers. It has become a significant information source for the higher education community. The Internet is not managed by any one organization. Rather it is a system of computer networks linked together in a cooperative, non-centralized collaboration.

The Internet developed from a research network created by the U. S. Department of Defense Advanced Research Projects Agency (ARPA) in 1969. The purpose of ARPANET was to connect various government, industrial, and research organizations. In 1990 ARPANET was dissolved and its functions were assumed by the federal government's National Science Foundation (NSF).

There are costs associated with setting up and maintaining an Internet connection. NSF and a number of corporate sponsors provide a majority of the funding. Organizations must pay an installation charge to connect to Internet and a yearly maintenance fee. Internet services are, however, "free" to the users. Vast amounts of information and computer services are free once you are connected to the Internet.

Internet can be accessed through companies that provide computer information services such as DELPHI (see Appendix D). Most universities in North America have Internet connections. Internet can usually be accessed directly through a central university computer. A better method is to have a microcomputer that is connected directly to Ethernet (this requires separate hardware). In this way you can have direct access to Internet services and download files to your computer.

The capabilities of Internet can be grouped into four major areas: (1) FTP, or File Transfer Protocol, a means of exchanging files between host computers and downloading files from a foreign host; (2) Telnet, a remote log-on procedure for accessing programs on remote computers as though they were local; (3) E-mail, an electronic mail service that allows the exchange of mail messages between Internet users and many networks outside of the Internet; and (4) Newsgroups and discussion lists that disseminate information to groups of users that provide a forum for researchers.

FTP accounts for almost half of the Internet usage. The FTP application provides access to software, graphic images (including weather satellite images), and documents of interests to educators, researchers, and students. It is capable of transferring files between computers with different operating systems. File transfer is accomplished by allowing users to access remote computers through a guest account called *anonymous*. The password for these accounts is the e-mail address of the user. Anonymous login provides restricted access to public directories on remote computers. Once connected in this way, basic UNIX commands can be used to get directory listings, change directories, and transfer files.

To initiate an FTP connection, the computer being used must have TCP/IP or be connected to another local computer that is running this software. TCP/IP (Transmission Control Protocol/ Internet Protocol) is the set of instructions that determines the timing and sequencing of data exchange between computers in two locations. It allows computers to exchange "packets" of information regardless of the operating system or hardware. These packets could be part of an e-mail message, a program, or a data file.

Telnet is a software system that facilitates network communications. It is most often associated with a program that is distributed by the National Center for Supercomputer Applications at the University of Illinois at Urbana – Champaign. Telnet programs are available for different operating systems, including Macintosh, DOS, and Windows. Telnet is essentially a terminal emulation program that is used to initiate contact with FTP sites.

Each Telnet program establishes a connection with the remote computer somewhat differently. With the Macintosh version, the "Open Connection" command in the File menu is used to request a connection. This presents a dialog in which the FTP address of the remote computer can be given and an FTP session can be established (Step 3, Figure C.1).

A window is then opened that acts as a terminal emulator. A log-in prompt may or may not appear. The commands to log-on to the remote computer are:

```
user anonymous
<your e-mail address>
```

Commands are issued in the command-line format of UNIX. For example, the "ls" command provides a listing of files. The "dir" command issues a more detailed directory listing. The "cd <sub-directory>" command is used to change directories. "Binary" causes files transfers to be done in a binary format. The "get <fn>" transfers a file from the remote computer to the local computer. The basic UNIX commands for FTP are

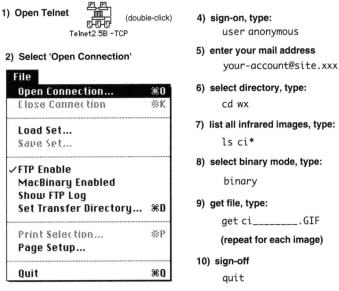

1) **Open Telnet**   (double-click)
   Telnet2.5B -TCP

2) **Select 'Open Connection'**

File

Open Connection...	⌘O
Close Connection	⌘K
Load Set...	
Save Set...	
✓**FTP Enable**	
MacBinary Enabled	
Show FTP Log	
Set Transfer Directory...	⌘D
Print Selection...	⌘P
Page Setup...	
Quit	⌘Q

3) **Specify FTP site, select 'FTP session'**

4) **sign-on, type:**
   user anonymous

5) **enter your mail address**
   your-account@site.xxx

6) **select directory, type:**
   cd wx

7) **list all infrared images, type:**
   ls ci*

8) **select binary mode, type:**
   binary

9) **get file, type:**
   get ci_____.GIF

   (repeat for each image)

10) **sign-off**
   quit

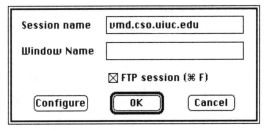

Session name	vmd.cso.uiuc.edu
Window Name	

☒ **FTP session (⌘ F)**

[Configure]   [OK]   [Cancel]

Figure C.1   Establishing an FTP connection with a remote computer. The Telnet program (NCSA) is used to establish a connection with a remote computer. Following this are the series of commands to transfer files from the remote computer.

summarized below:

ls	list files in the current directory
dir	more detailed file listing than ls
cd	change directory, e.g., "cd wx" changes to the wx directory
binary	sets to binary mode. Use when retrieving non-text files.
ascii	sets to asci mode (the default). Use for retrieving text files.
get	retrieves a file, e.g., "get readme" gets the file called readme
mget	retrieves multiple files, e.g., "get *.gif" retrieves all GIF files
bye	exits FTP

Figure C.2 is a listing of a session that retrieves a weather image from an FTP site.

```
220-FTPSERV2 at vmd.cso.uiuc.edu, 10:54:59 CDT TUESDAY 08/03/93
220 Connection will close if idle for more than 5 minutes.
user anonymous
331 Send your local userid instead of password for identification, please
<your e-mail address>
230-ANONYMOU logged in; working directory = ANONYMOU 191
cd wx
250 Working directory is WX 191 (ReadOnly)
ls ci*
200 Port request OK.
125 List started OK
CIR.GIF
CI080122.GIF
CI080200.GIF
.
CI080316.GIF
CI080317.GIF
250 List completed successfully.
binary
200 Representation type is IMAGE.
get CI080317.GIF
200 Port request OK.
150 Sending file 'CI080317.GIF'
138475 bytes received.
250 Transfer completed successfully.
quit
```

Figure C.2   A session with a remote FTP site. Commands that are used include **cd** (change directory), **ls** (list directory), **binary** (change to binary file transfer mode), **get** (to move a file to a local computer) and **quit** (to end the session). A series of files can be transferred by using the **mget** command.

There is a wealth of information available through Internet. The initial encounter with networks such as the Internet can be intimidating. Most people are not familiar with the UNIX-type commands that are required to transfer files. A certain amount of practice is needed to become accustomed to the procedure. Advanced topics — such as file compression and decompression, file types, and special searching utilities — are other areas that need to be mastered after one has become more familiar with Internet.

One of the major problems with Internet is finding a desired file. The available data sites on FTP changes almost daily. The service called "gopher" is a valuable tool to find these documents. Gopher is a protocol for "tunneling" through the Internet. A number of sites have also been established to index the tens of thousands of files that are available through anonymous FTP. These sites are called Archie servers and you will need to find the most accessible server for your area. In Telnet, login as "archie" (no password is needed) and type "prog filename" to find the location of the file. To search only for Macintosh software, add the suffix ".*\.hqx" to the end of the filename.

To access Internet services through e-mail when Internet is not available, send a message to "mail-server@rtfm.mit.edu" and insert the line "send usenet/news.answers/finding-sources" in the body of the message. A file will be e–mailed containing instructions on how to use FTP sites with e-mail. The University of Michigan also maintains an e-mail server for its Macintosh archive. Send a message to mac@archive.umich.edu with the word "help" in the body and instructions will be sent by e-mail.

An improved method of gaining access to Internet services is through NCSA Mosaic, an Internet-based global hypermedia browser for retrieving, displaying documents, and data from all over the Internet. Mosaic is part of the World-Wide Web project, a distributed hypermedia environment created through a collaboration by an international design and development team. Global hypermedia means that information located around the world is interconnected in an environment that allows you to travel through the information by clicking on hyperlinks — terms, icons, or images in documents that point to other, related documents. Any hyperlink can point to any document anywhere on the Internet. The program retrieves images and sound-clips by clicking on icons that are imbedded within text documents. Mosaic helps you explore a huge and rapidly expanding universe of information and gives you powerful new capabilities for interacting with information. Mosaic sites are prefixed with *http*.

The user should be aware of network etiquette. Kehoe, author of *Zen and the Art of the Internet: A Beginner's Guide* (available online), states that the user of anonymous FTP should be restricted to off-peak hours only. Courtesy requires that FTP logins not be attempted during normal business hours for the remote computer, especially from sites outside the United States. Large file transfers affect other people using the system because the computer is doing multiple jobs. Internet should be used responsibly.

What follows is a listing of the major Internet resources with their addresses. Many of these sites have FAQ (Frequently Asked Questions) or README files. Transfer these files first and read them before transferring other files.

MACINTOSH SOFTWARE

The five major Internet archives for shareware, freeware, and demo software are:

sumex-aim.stanford.edu (36.44.0.6)	Stanford University
mac.archive.umich.edu (141.211.165.41)	University of Michigan
ftp.apple.com (130.43.2.3.)	Apple Computer
ftp.support.apple.com	Apple Computer
wuarchive.wustl.edu (128.252.135.4)	Washington University

The last of these, wuarchive, is a mirror site that holds a copy of the software at the other sites. It often holds onto files after they have been removed from other sites for space considerations. The 'ftp.support.apple.com' site is Apple's semi-official repository for system software, developer tools, source code, and technical notes.

Macintosh files that are available through the Internet are compressed to take up less space and modified so that they can reside under a UNIX file system. This suffix indicates how a file has been compressed and modified. A file called *filename.xxx.yyy* has a compression method *xxx* and is modified by *yyy*. The most common Macintosh file compression program is StuffIt, indicated by a suffix *sit*. There are two file modification techniques, BinHex and MacBinary, and their suffixes are *hqx* and *bin*. A freeware program called StuffIt Expander will convert most desired file formats used at Macintosh ftp sites.

WINDOWS SOFTWARE

The most common way for PCs to share files over the network is through an archive. An archive is a file containing several other files that have been compressed in some fashion to save disk space. There are different types of archives, such as .ARC, .ZOO, and .ZIP, each compressed with a different archiving program. Programs are available to extract these files. For example, to extract a file in .ZOO format, the following command would be used with the ZOO program:

ZOO -extract <archive filename>

An individual filename can be specified after the name of the archive to extract a particular file. The programs ZIP and UNZIP work in a similar way. The major FTP sites for Windows software are "wuarchive.wust1.edu" at the University of Washington and "ftp.cica.indiana.edu" at the University of Indiana. The latter site contains over 300 MB of public domain and shareware files.

MAP FILES

There are a number of sites that store map files in vector format and programs that manipulate these files. The sites are:

alum.wr.usgs.gov[128.128.19.24] —> Digitized data base (lat/long) of geologic faults and earthquake epicenters.

charon.er.usgs.gov —> Software for UNIX systems for the plotting of maps and map projections. Files in Arc/INFO export format are also available. All files are in *tar* format.

csn.org —> Mapping software and data sets (/pub directory).

dis2qvarsa.er.usgs.gov —> Files in Arc/INFO export format. A map file that contains the outlines of the counties for all fifty states is available here in the /states100 directory.

edcftp.cr.usgs.gov —> The General Cartographic Transformation Package (GCTP) in the C programming language.

enterprise.nwi.fws.gov —> 13,000 national wetland inventory maps.

fermi.jhuapl.edu —> Sixty color shaded relief maps for most of the continental U.S.

ftp.cs.toronto.edu —> Spherical coordinates of major U.S. cities and the directory of the CIA world map database.

ftp.uu.net —> Digitized outline of the U.S. and county boundaries.

gatekeeper.dec.com —> Over 30 city, state, and world PostScript maps (pub/maps). Also stores the CIA World Data Bank.

greenwood.cr.usgs.gov —> Digital geology maps in Arc/Info format.

hanauma.stanford.edu —> CIA World Data Bank II, a large data file of coastlines, countries, rivers, islands, and lakes of the world.

http://www.delorme.com —> Sample maps available from downloading from this commercial firm. (Mosaic)

moon.cecer.army.mil —> GRASS GIS source code and sample database.

ncardata.ucar.edu and **www.ucar.edu** (Mosaic) —> Digital elevation models.

spectrum.xerox.com —> Digital Line Graph (DLG), Digital Elevation Model (DEM) TIGER map files. The file README-MAP in the pub/map directory describes the data in the sub-directories dlg, dem, and tiger. The pub/map/dem subdirectory contains over 60 digital elevation models. The tiger directory contains files for nine selected cities in the United States.

stardent.arc.nasa.gov —> Digital Elevation Model data from Mars sampled at a variety of spatial resolutions.

walrus.wr.usgs.gov —> A world elevation data set.

IMAGE FILES

The most common type of graphic data available on the Internet are images from meteorological satellites. These are raster images that are usually stored in the GIF format. Compression techniques may have also been applied to these files.

The files are not in a raw format — they have all been modified by computers on the ground. For example, the weather images contain state outlines that have been added to the image. Some of the weather images are actually maps that depict weather fronts. Most of the images have been geometrically and/or radiometrically corrected.

The number of FTP sites store a variety of weather images. File names beginning with ci are infrared files; cv indicates the visual spectrum; sa files are surface analysis maps; and u files map the upper level pressures. These files are updated hourly. The sites with imagery include:

LANDSAT imagery	Directory
vab02.larc.nasa.gov	pub/gifs/misc/landsat

GOES satellite images of North America

ats.orst.edu	
early-bird.think.com	pub/weather/maps
ftp.colorado.edu	pub/weather-images
ftp.uwp.edu	pub/wx
kestrel.umd.edu	pub/wx
wmaps.aoc.nrao.edu	pub/wx
wuarchive.wust1.edu	multimedia/images/wx
wx.research.att.com	wx
westsat.com	pub/images

AVHRR satellite images

sanddunes.scd.ucar.edu	AVHRR 1 KM full scene passes, bands 1-5
aurelie.soest.hawaii.edu	AVHRR data for the area around Hawaii

SIR-C / X-SAR Radar Images
jplinfo.jpl.nasa.gov

Manned Space Images

sseop.jsc.nasa.gov	Space Shuttle imagery
139.169.21.11	Over 200,000 images from Mercury to Shuttle
ames.arc.nasa.gov	A variety of earth images from space

Other sites with satellite images include:

cumulus.met.ed.ac.uk	Meteosat images of Europe
ftp.fu-berlin.de	Meteosat images of Europe and N. Atlantic
marlin.jcu.edu.au	GMS-4 images of Australia
spot.colorado.edu	surface and radar maps of N. America
rainbow.physics.utoronto.ca	NOAA-11/12 images of N. America
explorer.arc.nasa.gov	Viking, Magellan and Voyager images
snow.nohrsc.nws.gov	snow cover maps of US

SOCIO-ECONOMIC DATA

These files are produced by world, federal, or state government agencies. At the site cwis.usc.edu in the directory *Library and Research Information/Research Information/Government Information* are the following documents:

1. CIA World Factbook
2. Country Reports - Economic Policy and Trade Practices
3. 1990 Census (not recommended for FTP)
4. Economic Bulletin Board
5. Economic Conversion Information Exchange Database
6. National Geophysical Data Center
7. U.S. Department of Agriculture Data
8. U.S. EPA National Online Library System
9. U.S. Industrial Outlook
10. U.S. Budget
11. United Nations Data

The site **ftp.cs.toronto.edu** also maintains the CIA World Factbook in *doc/geography*.

The United States Bureau of the Census has an information server on the Internet. The service can be accessed in one of three ways:

ftp://ftp.census.gov/pub	# use with ftp
http://www.census.gov/	# use with mosaic, lynx, etc
gopher://gopher.census.gov	# use with gopher

If you have problems, questions, suggestions about this server, send e-mail to: gatekeeper@census.gov

FAQ FILES

FAQ files contain updated information on the location of files, subscribing to mailing lists, etc. These files can be obtained through ftp.

Computer Graphics: ftp:// rtfm.mit.edu/pub/usenet/news.answers
Data Formats: ftp:// rtfm.mit.edu/pub/usenet/news.answers/sci-data-formats
GIS:
    ftp:// abraxas.adelphi.edu/pub/gis/FAQ
    ftp:// ftp.census.gov.geo/pub/geo/gis-faq.txt
    ftp:// gis.queensu.ca/pub/gis/docs/gissites.txt
    ftp:// ncgia.ucsb.edu (various FAQ's and papers)
GIS and Remote Sensing: ftp:// ftp.gis.queensu.ca/pub/gis/docs

Image Processing:
  ftp:// zippy.nimh.nih.gov/pub/nih-image
  ftp:// rtfm.mit.edu /pub/usenet-by-hierarchy/news/answers
  ftp:// ruby.oce.orst.edu /pub/sci.image.processing
  ftp:// boombox.micro.umn.edu /pub/gopher/Macintosh-TurboGopher
  ftp:// ftp.soils.umn.edu:pub/info/email-lists/Nih-image
Internet: ftp:// quartz.rutgers.edu/pub/internet/docs.
Meteorology: ftp:// rtfm.mit.edu/weather/data/part1 (also part2)
Resources for the Earth Sciences: ftp:// ftp.csn.org.ores.txt
World-Wide Web: ftp:// ftp.eit.com/pub/web.guide/guide61.txt

INTERNET SOURCES

The Whole Earth Internet User's Guide & Catalog, by Ed Krol (ISBN 1-55558-033-5). Discusses how Internet works and how to use it. Available from: O'Reilly & Associates, Inc., 103 Morris Street, Stuite A. Sebastopol, CA 95472; (800) 998-9938 or (707) 829-0515.

Zen and the Art of the Internet: A Beginner's Guide, by Brendan P. Kehoe (ISBN 0-13-010778-6). An introduction for the novice Internet user. Covers basic Internet features such as FTP, Telnet, electronic mail, and newsgroups. Available from: PTR Prentice Hall, 113 Sylvan Avenue, Route 9W, Englewood Cliffs, NJ 07632; (201) 592-2348. Also available from ftp:// quartz.rutgers.edu/pub/internet/docs.

The Internet Companion: A Beginner's Guide to Global Networking, by Tracy LaQuey and Jeanne C. Ryer (ISBN 0-0201-62224-6). A concise guide to Internet basics; e-mail; FTP; finding out about Archie and Gopher; and descriptions of other Internet tools and resources. Available from: Addison-Wesley, One Jacob Way, Reading, MA 01867; (617) 944-3700.

OTHER REFERENCES

DERN, D. P. (1994) *The Internet Guide for New Users*. New York: McGraw-Hill.

FRAASE, M. (1993) *The Mac Internet Tour Guide*. Chapel Hill, N.C.: Ventana Press.

HARDIE, T.L. AND NEOU, V. (eds.). (1993) *Internet Mailing Lists*. Englewood Cliffs, NJ: Prentice Hall.

LAQUEY, T. (1993) *The Internet Companion*. Reading, MA: Addison-Wesley.

MALAMUD, C. (1993) *Exploring the Internet*. Englewood Cliffs, NJ: PTR Prentice Hall.

MARIE, A., KIRKPATRICK, S., NEOU, V., AND WARD, C. (1993) *Internet: Getting Started*. Englewood Cliffs, NJ: PTR Prentice Hall.

Appendix D

Other Sources for Software

Programs for interactive and animated cartography are in a constant state of development. The only way to distribute an up-to-date list of the software is through Internet, because the files can be constantly updated. To obtain a listing of the available software and example cartograhic animations, use the Mosaic software to access the site **gopher://rsal.unomaha.edu**. Files are kept in a folder called **animated cartography**.

Other sources of software include user groups, individuals, companies that distribute software at reduced prices and computer information services. The following listing includes these other sources.

MICROCOMPUTER USER GROUPS

AAG Microcomputer Specialty Group
IBM Software Exchange
Department of Geography
Indiana University of Pennsylvania
Indiana, PA 15705
Tel. (415) 357-2251
FAX: (412) 357-3768
Comments: A number of mapping related programs are distributed through this software exchange. Programs include MicroCAM and the MicroCAM Interface software. MicroCAM produces map projections of the world. The MicroCAM interface program generates MicroCAM command files.
AAG Microcomputer Specialty Group

Macintosh Software Exchange
Paul Sinasohn
3359 Arkansas Street
Oakland, CA 94602
E-mail: sinasohn@netcom.com
Comments: A number of programs and HyperCard stacks are available from this
exchange.

PROGRAMS FROM AUTHORS

ExploreMap
Department of Geography
University of Kansas
Lawrence, KA 66045
Comments: An MS-DOS–based program for choropleth mapping called
EXPLOREMAP.

WORLD PROGRAM
25 Blegen Hall
University of Minnesota
269 19th Ave. S.
Minneapolis, MN 55455
Comments: An MS-DOS–based program that produces over 150 different map
projections.

SOFTWARE DISTRIBUTION SERVICES

Chariot Software Group
3659 India St., Suite 100c
San Diego, CA 92103
Tel.: (800) 242-7468
FAX: (800) 800-4540
Comments: Distributes educational software for the Macintosh, including map files and
programs for mapping.

Douglas-Stewart
2402 Advance Rd.
Madison, WI 53704
Tel: (800) 279-2003
Comments: Distributes the major software packages for Macintosh and Windows at
reduced prices to educational institutions.
Educorp Computer Services
7434 Trade Street

San Diego, CA 92121-2410
Tel.: (619) 536-9999
FAX: (619) 536-2345
Comments: Primarily distributes CD-ROM, HyperCard stacks, and utility programs for
the educational market.

COMPUTER INFORMATION SERVICES

All of these services provide access through a telephone and a modem to a large data
base. Files includes weather images, demographic data files, programs, maps as well as
airline schedules and other information.

America Online
8619 Westwood Center Drive
Vienna, VA 22182-9806
Tel.: (800) 827-6364

Compuserve
5000 Arlington Centre Blvd.
Columbus, OH 43220
Tel.: (617) 457-8600

DELPHI Internet Services
1030 Massachussets Ave.
Cambridge, MA 02138-5302
Tel: (800) 695-4005
e-mail: INFO@delphi.com

Genie
GE Information Services; Dept. 02B
401 North Washington St.
Rockville, MD 20850
Tel.: (800) 638-9636

DISCUSSION LISTS:

The following are discussion lists that deal with specific issues related to cartography. Mail messages are posted and then distributed to subscribers. The purpose of these lists is to share ideas and experiences. The discussion often concerns specific software products. To subscribe to a discussion list send a mail message to the address provided below. Type SUBSCRIBE *name-of-discussion-list* and your first and last name in the body of the message.

Discussion List	Topic	Subscription Address
esri.com	ESRI software products	listserv@esri.com
GIS-L	GIS related issues	listserv@ubvm.cc.buffalo.edu
UIGIS-L	GIS user interface issues	listserv@ubvm.cc.buffalo.edu
GPS Digest	Global Positioning System	gps-request@tws4.si.com
IDRISI-L	IDRISI Raster GIS	mailserv@toe.towson.edu
Ingr-En	Intergraph software products	mailserv@ccsun.tuke.sk
MAPINFO-L	MapInfo software products	majordomo@csn.org
TGIS-L	Temporal issues in GIS	listserv@ubvm.cc.buffalo.edu

Appendix E

Sources of Maps and Imagery

MAP SOURCES

Central Intelligence Agency
Attn: Public Affairs
Washington, DC 20505
Phone: (703) 351-2053
Products: World and regional maps.

Earth Science Information Center
U.S. Geological Survey
507 National Center
Reston, VA 22092
Phone: (800) 872-6277;
(703) 860-6045
Comments: Provides information
about maps distributed through the
USGS.

National Cartographic Information
Center
U.S. Geological Survey
507 National Center
Reston, VA 22092
Comments: Provides information on
available cartographic data..

National Geographic Society
P. O. Box 1640
Washington, D.C. 20013-1640
Phone: (800) 447-0647
Products: Maps and atlases of the
world.

National Geophysical Data Center
NOAA, Code E/GC1
325 Broadway
Boulder, CO 80303
Phone: (303) 497-6764
FAX: (303) 497-6513
E-mail: info@ngdc1.colorado.edu
Products: Digital topography data

USGS Map Sales
P.O. Box 25286
Denver, CO 80225
Phone: (303) 236-7477
Products: U.S. and World maps
produced by the USGS.

IMAGERY SOURCES

EROS Data Center
Sioux Falls, SD 57198
Products: NASA and NHAP black
and white and color photography.

EOSAT
North American Sales
6000 Lombardo Center Drive, Ste 310
Cleveland, OH 44131
Products: LANDSAT satellite
imagery.

Aerial Photography Field Office
ASCS-USDA
P. O. Box 30010
Salt Lake City, UT 84125
Products: Large-scale, black & white
U.S.D.A. Agricultural Survey and
Conservation Service photography.

Chicago Aerial Survey, Inc.
2140 Wolf Road
Des Plaines, IL 60018
Products: Large-format camera
photographs.

EOSAT Corporation
4300 Forbes Blvd.
Lanham, MD 20706
Products: LANDSAT images.

SPOT Image Corporation
1897 Preston White Dr.
Reston, VA 22091-9949
Phone: (800) 275-7768
Products: SPOT satellite imagery in
digital and photographic formats.

Satellite Data Service Branch
NOAA/National Environmental
Satellite, Data, and Information
Service
World Weather Building, Room 100
Washington, DC 20233
Products: Weather images.

Index